DELEUZE AND GUATTARI FOR ARCHITECTS

Thinkers for Architects

Architects have often looked to philosophers and theorists from beyond the discipline for design inspiration or in search of a critical framework for practice. This original series offers quick, clear introductions to key thinkers who have written about architecture and whose work can yield insights for designers.

Deleuze and Guattari for Architects

Andrew Ballantyne

Heidegger for Architects

Adam Sharr

Irigaray for Architects

Peg Rawes

THINKERS FOR ARCHITECTS

Deleuze and Guattari

for

Architects

Andrew Ballantyne

LONDON AND NEW YORK

First published 2007
by Routledge
2 Park Square, Milton Park, Abingdon, Oxon OX14 4RN

Simultaneously published in the USA and Canada
by Routledge
711 Thomas Avenue, New York, NY10017

Routledge is an imprint of the Taylor & Francis Group, an informa business

Typeset in Frutiger and Galliard by
Florence Production Ltd, Stoodleigh, Devon
Printed and bound in Great Britain by
TJ International Ltd, Padstow, Cornwall

British Library Cataloguing in Publication Data
A catalogue record for this book is available from the British Library

Library of Congress Cataloging in Publication Data
Ballantyne, Andrew.
Deleuze and Guattari for architects / Andrew Ballantyne.
p. cm. -- (Thinkers for architects series)
Includes bibliographical references and index.
1. Deleuze, Gilles, 1925–1995. 2. Guattari, Félix, 1930–1992.
3. Architecture – Philosophy. I. Title.
B2430.D454B36 2007
194 – dc22 2007013342

ISBN10: 0–415–42115–2 (hbk)
ISBN10: 0–415–42116–0 (pbk)
ISBN10: 0–203–93420–2 (ebk)

ISBN13: 978–0–415–42115–7 (hbk)
ISBN13: 978–0–415–42116–4 (pbk)
ISBN13: 978–0–203–93420–3 (ebk)

To Peter, Joanna and Penelope Klein

Contents

Series Editor's Preface

Adam Sharr

Architects have often looked to thinkers in philosophy and theory for design ideas, or in search of a critical framework for practice. Yet architects and students of architecture can struggle to navigate thinkers' writings. It can be daunting to approach original texts with little appreciation of their contexts and existing introductions seldom explore architectural material in any detail. This original series offers clear, quick and accurate introductions to key thinkers who have written about architecture. Each book summarizes what a thinker has to offer for architects. It locates their architectural thinking in the body of their work, introduces significant books and essays, helps decode terms and provides quick reference for further reading. If you find philosophical and theoretical writing about architecture difficult, or just don't know where to begin, this series will be indispensable.

Books in the *Thinkers for Architects* series come out of architecture. They pursue architectural modes of understanding, aiming to introduce a thinker to an architectural audience. Each thinker has a unique and distinctive ethos, and the structure of each book derives from the character at its focus. The thinkers explored are prodigious writers and any short introduction can only address a fraction of their work. Each author – an architect or an architectural critic – has focused on a selection of a thinker's writings which they judge most relevant to designers and interpreters of architecture. Inevitably, much will be left out. These books will be the first point of reference, rather than the last word, about a particular thinker for architects. It is hoped that they will encourage you to read further; offering an incentive to delve deeper into the original writings of a particular thinker.

The first three books in the series explore the work of: Gilles Deleuze and Félix Guattari; Martin Heidegger; and Luce Irigaray. Familiar cultural figures, these are thinkers whose writings have already influenced architectural designers and critics in distinctive and important ways. It is hoped that this series will expand over time to cover a rich diversity of contemporary thinkers who have something to say to architects.

Adam Sharr is Senior Lecturer at the Welsh School of Architecture, Cardiff University, and Principal of Adam Sharr Architects. He is author of *Heidegger's Hut* (MIT Press, 2006), *Heidegger for Architects* (Routledge, 2007), joint editor of *Primitive: Original Matters in Architecture* (Routledge, 2006) and Associate Editor of *arq: Architectural Research Quarterly* (Cambridge University Press).

Illustration Credits

Artists' Rights Society, page 39.

Andrew Ballantyne, page 74; page 75; page 93.

Columbia Pictures, page 24.

Damien Hirst, page 10.

Gerard Loughlin, page 77.

MGM, page 59.

Reuters/Benoit Tessier, page 96.

Uexküll, 1934, page 82.

Warner Brothers, page 68; page 69.

Acknowledgements

Thank you to the people who have encouraged me, and to those who have been patient with me while I was preoccupied writing. They include my colleagues in the Tectonic Cultures Research Group at Newcastle University, and Emily Apter, Dana Arnold, Steve Basson, Ed Dimendberg, Jean Hillier, Neil Leach, Gerard Loughlin, Erin Manning, Brian Massumi, Sally Jane Norman, John Paul Ricco, Anne and Joseph Rykwert, Adam Sharr, Chris Smith and Anthony Vidler.

Peter Klein introduced me to Deleuze and Guattari's work in London in 1982. We were in a bookshop, and there was a pile of remaindered copies of the first American edition of *Anti-Oedipus*, which he drew to my attention, astonished to see it there. Even at the knock-down price it seemed expensive for an impulse-buy.

'Is it good?' I asked.

'What can I say?' he said, '. . . It changed my life.'

Andrew Ballantyne
Asquins
1 January 2007

CHAPTER 1

Who?

No longer ourselves

Gilles Deleuze and Félix Guattari worked together on several books, and worked separately on many more. Their best known work stretched across two volumes with the title *Capitalism and Schizophrenia* – volume 1, *Anti-Oedipus* (1972); volume 2, *A Thousand Plateaus* (1980). Separately Deleuze (1925–95) was a professional philosopher, and Guattari (1930–92) was a psychiatrist and political activist. When they collaborated, their individual voices cannot be separated out and they seem to dissolve into one another. Sometimes the writing shifts into a new register as a *persona* is briefly adopted in order to give an impression of what the topic looks like from a particular point of view – but these points of view can seem bizarrely idiosyncratic – the point of view of a molecule, a moviegoer, or a sorcerer. 'The two of us wrote *Anti-Oedipus* together,' they said, 'Since each of us was several, there was already quite a crowd' (Deleuze and Guattari, 1980, 3). Personal identity here is something that is taken up, and then dropped or reformulated, so who were they really, these slippery characters? How would we say who they were? More importantly, why would we want to know? And if, at some point, we felt that we knew who they were, then what would it be that we would know? Their aim, they say, is 'to reach, not the point where one no longer says I, but the point where it is no longer of any importance whether one says I.' The question 'who?' simply will not arise;

. . . if, at some point, we felt that we knew who they were, then what would it be that we would know?

nevertheless for the time being they have kept their names 'out of habit, purely out of habit', but then disconcertingly they conclude: 'We are no longer ourselves.' Whatever people say they are, that's what they're not. Here, on the opening page of *A Thousand Plateaus*, is a succinct but determined challenge to our usual habits of thought, and it seems to derive from two principal sources: Guattari's work with psychiatric patients, and Deleuze's philosophical habits of mind, looking for rigorous logic while setting aside the common-sense expectations that would normally deflect us from following the logic through to its conclusions. There is often a role for common sense in our lives, and Deleuze and Guattari notice themselves using it for example when they signed their book with their own names. 'It's nice to talk like everyone else, to say that the sun rises, when everybody knows it's only a manner of speaking' (Deleuze and Guattari, 1980, 3).

It's nice to talk like everyone else, to say that the sun rises, when everybody knows it's only a manner of speaking.

Of course the sun rises – with our own eyes we can see it happening, if we go at dawn to a place with a distant horizon in the east. Nevertheless we know that the earth orbits the sun, and from a more sophisticated point of view the 'sunrise' is a very limited earth-bound description – pedestrian, commonplace, but often the most useful thing to say. How pedantic it would sound to insist on any other description in a normal social gathering. It might be exhilarating to sense oneself at that moment watching a static sun while the earth turned so as to allow a clearer view of it, riding Spaceship Earth, but probably that is something to do as a private act of the imagination. If the thought occurs to me while I'm standing in a queue at a bus stop, then it's not a thought I'm going to share with the person standing next to me. I would go for a commonplace remark about the sunrise. If a stranger turned to me and started talking about 'Spaceship Earth', then I would start to react, I think, by feeling anxious.

Character-defining questions

If I try to explain who Deleuze and Guattari were, then I start by trying to think about the character-defining things they did. And what they did – so far as their international audience is concerned – was to present new ways of conceptualizing things. There are other ways of saying who someone is.

John Berendt wrote the novel *Midnight in the Garden of Good and Evil*, set in Savannah, Georgia; and he put it very succinctly. According to one of his acquaintances in Savannah, 'If you go to Atlanta, the first question people will ask you is, "What's your business?" In Macon they ask, "Where do you go to church?" In Augusta they ask your grandmother's maiden name. But in Savannah the first question people ask you is "What would you like to drink?"' (Berendt, 1994, 30–31). The answers to these questions are identity-defining. If my grandmother is not someone who is known in Augusta then I too am nobody: I can buy things in the shops, and eat in the restaurants, but it is to be expected that I will never fully establish myself personally as part of that society, but if I have grandchildren then they might make it.

. . . in Savannah the first question people ask you is 'What would you like to drink?'

If I go to Atlanta without any business to declare, then again apparently I am nobody (even if my grandmother was born there). Even in Georgia things are not so clearly defined that these rules would always hold. However the answers hardly matter: the important point for the story is that the questions themselves define the identities of the places where they are asked. Atlanta is *nouveau-riche*, Augusta is snobbish, Savannah is hedonistic; or so we might suppose from the characterization. This is how one's identity is determined, and equally how we disappear from view if we cannot lay claim to an identity that is recognizable. However it is not only in different places, or in different historical epochs that different identity-defining questions come to the fore. Genealogies are identity-defining in aristocratic societies with hereditary titles and roles – for princes and the nobility of course, even today – but even the librarians' posts at Versailles were hereditary, and much lower down the social scale there was often something similar but less legalistic going on. In a very stable society that does not change from one generation to the next, for reasons that feel more practical than ideological, the person best placed to learn the skills of a shoemaker or a joiner might be the craftsman's son, who had access to the workshop, and the most complete trust of the owner of the business, his father. A skilled artisan's son would be the person most likely to succeed him in his business. So the boy's parentage would seem to be an important and character-defining thing about him. In the twenty-first century there is more spatial and social mobility than there was even 50 years ago, and the tracing of personal genealogies has never been more popular. We feel, when we find out something about our forebears, that we have learnt something about ourselves. Even when we have thoroughly uprooted ourselves and are working in places that our relatives do not know, and in ways that they do not understand, personal genealogy reasserts itself on family occasions.

Both identities are real. They are both roles that she knows how to play.

A woman who runs an international company and has hundreds of employees to do her bidding at the office, is redescribed for the family occasion as somebody's daughter, or somebody's aunt, and that is her identity for the duration. Both identities are real. They are both roles that she knows how to play. We have different ways of saying who someone is, and the way that we use will depend on the company we're in, or on the occasion. So it is correct to say, for example, that Gilles Deleuze was the husband of Fanny, and the father of Julien and Emilie, but what is that to us? It sounds overly gossipy even to have mentioned it. It would be correct to say that Deleuze was a good tennis player, and a bad driver, but these details are unimportant to us now that no one will be in a position to play tennis with him, or politely to decline the offer of a lift. There is a tendency in biographical writing to suppose that when we see the subject off guard, intimately, perhaps behaving badly, then we see the person in their truest light, as if there is an innermost identity that is really and truly our personal identity when all the public identities have fallen away, and

Identity is political, in that it is generated through our relations with others.

which we would do our best to keep hidden. Deleuze and Guattari resist that idea. Identity is political, in that it is generated through our relations with others.

It is not altogether interior, but has an external aspect. Our various temporary identities are all the identities we have, and depending on the point that we are addressing, the pertinent identity is the one – or maybe more than one – that has a bearing on the case. So if we are reading Deleuze's philosophy, it is beside the point how well he drove his car or looked after his nails.[1] And if I try to explain who Deleuze and Guattari were, then I cannot succinctly explain what was their innermost essence, and move on to other matters. What I have to do is to say what they did, and one of the things that they did was to make the idea of identity problematic. They were *by definition* the people who did those things – that is their identity for our purposes. And so far as I am concerned, what is interesting about them are the ideas that they formulated and wrote down. Their identity here is as authors of texts and creators of concepts, and it will assemble itself gradually as we see something of those texts and concepts below.

. . . for the kind of architect who wants to be stimulated into extending the range of what life has to offer, Deleuze and Guattari's attitudes will immediately be congenial.

These texts and concepts are never an end in themselves. They are deliberately experimental, and the point of them is always to see what might be turned up that could bring about new possibilities in living. In this stance we see that there is a link with a certain sort of architect – the sort who wants to design buildings that promote life and that are experiments in living. There are other sorts of architects, and other sorts of thinkers, who would adopt a different approach, and they will find Deleuze and Guattari's writings unappealing; but for the kind of architect who wants to be stimulated into extending the range of what life has to offer, Deleuze and Guattari's attitudes will immediately be congenial – even if it may take a little longer to make sense of their concepts.

Lines of flight

Part of the problem that one faces in trying to write about the things that really matter is that we have to be in one state of mind to experience the things that matter, and in a completely different frame of mind in order to say something about that experience in words, and to write the words down. For example with an aesthetic experience, the thrill of being moved by a work of art – whether it be literary, architectural or whatever – is difficult to reproduce in a commentary. Perhaps it works most powerfully and effectively when one is taken by surprise, and the rush of the sublime overtakes one unannounced. If I am visiting someone's dwelling, then I expect to find there the means of sustaining a life and would think it a successful home if it did no more than that. If I have travelled a thousand miles to see an architectural marvel then I will be disappointed if I don't find something more – or more accurately: something else. I want to feel the earth move beneath my feet. I want a sense that a view of the heavens has opened up to me; or that the universe has rearranged itself around me – something of that order – a sense of the oceanic, or at least vertigo, standing on a precipice of history, or suddenly becoming aware of a horizon that was there all along, behind the things that were close at hand in everyday life. My experience of the sublime is real enough, but just saying that doesn't make you feel it – I have to say other things that rekindle the feeling in me, and then hope for the best.

. . . a horizon that was there all along, behind the things that were close at hand in everyday life.

Even if I have felt thrilled by a particular building, there is no guarantee that you will feel the same way if I take you to the same place. And certainly it was absent in the instructions that were given to the builders: a wall here, 3 metres high, 80cm thick; a glass panel there, held in place by a neoprene seal. There is no magic in the instructions. Similarly with literature, the words on the page might describe one thing, but the effect that matters is produced in an altogether different register. For example, Scott Fitzgerald wrote intelligently and perceptively, but what makes him a great writer was the fact that through the descriptions of gilded lives (swooning in the heat, running on alcohol) a turn of phrase here, and a detail of action there, opens up a sense of hollowness within, and with a rush the fast cars and the beautiful clothes become mere furniture in the acting out of something as elemental as a Greek tragedy, and the tawdry glamour of his protagonists' lives is infused with some sort of epic grandeur. This 'transport', this 'lift', is, to use Deleuze and Guattari's term, produced by a 'line of flight'. Works of art produce lines of flight. A warm bath is a pleasant experience, but I normally experience it as a comfort, rather than as a work of art. Many paintings, films and novels work in this sort of way, like a warm bath, producing a sense of good-natured well-being, rather than the rush of air, the enthusiasm, of a line of flight. There is a role for such experiences in my life, and I would not wish them away so as to live exclusively on the greatest art and bracing showers. Always to choose the comfortable over the sublime would be a formula for mediocrity, but conversely the asceticism of the determined artist can turn out to have been a formula for mania, alcoholism or anorexia (Deleuze and Parnet, 1977, 50–1).

> **A flight is a sort of delirum. To be delirious is exactly to go off the rails (as in *déconner* – to say absurd things, etc.). There is something demonaical or demonic in a line of flight. Demons are different from gods, because gods have fixed attributes, properties and functions, territories and codes: they have to do with rails, boundaries and surveys. What demons do is jump across intervals, and from one interval to another. (Deleuze and Parnet, 1977, 40)**

. . . the asceticism of the determined artist can turn out to have been a formula for mania, alcoholism or anorexia.

Deleuze evokes this mood when describing the work of Baruch Spinoza (1632–77) especially the *Ethics*, which remained unpublished in Spinoza's lifetime, for fear of the consequences for his personal safety:

> **Many commentators have loved Spinoza sufficiently to invoke a Wind when speaking of him. And in fact no other comparison is adequate. But should we think of the great calm wind the philosopher [Victor] Delbos speaks of? Or should we think of the whirlwind, the witches' wind spoken of by 'the man from Kiev,' a non-philosopher par excellence, a poor Jew who bought the *Ethics* for a kopek and did not understand how everything fitted together. (Deleuze, 1970, 130)[2]**

This 'man from Kiev' is a fictional character in Bernard Malamud's novel *The Fixer,* who said of Spinoza's *Ethics*, 'I read through a few pages and kept on going as though there were a whirlwind at my back. As I say, I didn't understand every word but when you're dealing with such ideas you feel as though you were taking a witches' ride. After that I wasn't the same man.' (Malamud, 1966; quoted by Deleuze, 1970, 1). The sense of delirious flight is palpable. By redescribing everything, Spinoza awakens the sense of alternative possibilities, and destabilizes the common-sense order of things. 'If we are Spinozists,' says Deleuze:

> **we will not define a thing by its form, nor by its organs and functions, nor as a substance or a subject. Borrowing terms from the Middle Ages, or from geography, we will define it by *longitude* and *latitude*. A body can be anything: it can be an animal, a body of sounds, a mind or an idea; it can be a linguistic corpus, a social body, a collectivity. We call longitude of a body the set of relations of speed and slowness, of motion and rest, between particles that compose it from this point of view, that is between *unformed elements*. We call latitude the set of affects that occupy a body at each moment, that is, the intensive states of an *anonymous force* (for existing, capacity for being affected). In this way we construct the map of a body. The longitudes and latitudes together constitute Nature, the plane of immanence or consistency, which is always variable and is constantly being altered, composed and recomposed, by individuals and collectivities. (Deleuze, 1970, 127–8)**

A body can be anything: it can be an animal, a body of sounds, a mind or an idea; it can be a linguistic corpus, a social body, a collectivity.

Through the fog of unfamiliar terminology here, there is a sense of the pervasive provisionality, seeing the world as an open flux of possibilities, that makes Deleuze and Guattari's writings so appealing to architects, who find themselves called upon to find form for buildings. A precondition for finding form is to be without form, to suspend the condition of having form, so that a new possibility can emerge. If we define a building type by its form, then it does not need a designer – we already have the design – and in a non-progressive culture that could be how things are done. Most of the time in order to function well in a culture, we need to be able to deal with common sense, and to be able to see

What they can help us to do is to keep common sense at bay.

things in commonplace ways, for example in order to give and receive meaningful instructions, to arrive at appointments on time, and to be trusted with the large sums of money that putting up buildings inevitably involves. Deleuze and Guatttari are no use for any of that. What they can help us to do is to keep common sense at bay, take us on a witches' ride, so that we can see the world as unformed elements acted upon by anonymous forces, and by entering into this world of virtualities, it can put us in a position where we might actualize something that has not previously been.

Away from the flock

On the spectacularly beautiful but sometimes bleak Cumbrian fells in the north of England, there are sheep which have lived in the same territory for countless generations. They know their way around. They do not wander away. They follow their habitual paths, the knowledge of which is passed on from one generation to the next. They are said to be 'hefted' – a word of Norse origin, now used only in this context. Hefted sheep are a boon for the region's hill farmers, because they do not need to be kept in walled enclosures. Their movements are predictable; they can be herded by dogs for dipping and shearing, but they are closer to being wild animals than are the sort of farm animals that need to be kept in special buildings, and can be allowed to graze only in enclosed fields. Fields here are enclosed by drystone walls, built by hard labour. The hefted sheep can be given complete freedom, because they do not think of making use of it. They are not going to wander off to the big city, into an art gallery, or even on to a main road (see p. 10). There are real dangers in this countryside, such as hillsides of shifting scree and cliffs with raging waterfalls. A free-spirited sheep could come to a sudden end here. But the hefted sheep know where they are supposed to be, and behave as though that is just what they want. They are bonded with specific pastures. They are

Damien Hirst, *Away From the Flock*, 1994.

'territorialized'. The hefted sheep live in a world that is governed entirely and uncomplicatedly by common sense. The received wisdom of the generations keeps them safe. If ever a sheep with a philosophical sensibility were born, the others would see it as mad, bad, and dangerous to know, and one way or another it would not last for long as part of the flock.

'Men believe they are free,' said Spinoza, 'precisely because they are conscious of their volitions and desires; yet concerning the causes that have determined them to desire and will they have not the faintest idea' (Spinoza, 1677b, 57). In fact they are so very securely in the grip of these desires, that they take them for granted and they believe themselves to have no control over them. The established codes of behaviour are kept in place by a strange complicity between tyrants and slaves, which the slaves accept and enforce:

'Men believe they are free,' said Spinoza, 'precisely because they are conscious of their volitions and desires; yet concerning the causes that have determined them to desire and will they have not the faintest idea'.

> **In despotic statecraft, the supreme and essential mystery is to hoodwink the subjects, and to mask the fear, which keeps them down, with the specious garb of religion, so that men may fight as bravely for slavery as for safety, and count it not shame but highest honour to risk their blood and lives for the vainglory of a tyrant. (Spinoza, 1677a, quoted by Deleuze, 1970, 25)**

Later the study of this issue was taken up by the German romantic philosopher Friedrich Nietzsche (1844–1900) especially in *Beyond Good and Evil*, and its sequel *The Genealogy of Morals* (its opening phrase: 'We are unknown to ourselves'). And we should note that not only did Deleuze write a book about Nietzsche (indeed he wrote two) but he made Nietzsche's name the opening word of his second book on Spinoza. The ideas would be taken much further in the first of Deleuze and Guattari's collaborations, *Anti-Oedipus*, as we will see in chapter 2, but of course it should not go unnoticed (how could it?) that the figure who particularly dominated the study of human desires and how they take shape is Sigmund Freud (1856–1939) who had immense prestige in French intellectual circles when *Anti-Oedipus* was published in 1972. Deleuze and Guattari's book title makes it clear that they had issues with him, as it was he who proposed a set of unconscious desires, which he called *'the Oedipus complex'*, as the unacknowledged basis for many of our actions. However another of Deleuze's formative enthusiasms should be mentioned at this point: the Scottish philosopher, David Hume (1711–76).

Backgammon

Hume was the subject of Deleuze's first book, and it is his conception of the 'self' that is illuminating here. For Hume the self was a very provisional concept, and when he thought carefully about it in a rigorous and philosophical way he could not persuade himself that he had any good reason for believing in it. But:

> **Most fortunately it happens, that since reason is incapable of dispelling these clouds, nature herself suffices to that purpose, and cures me of this philosophical melancholy and delirium, either by relaxing this bent of mind, or by some avocation, and lively impression of my senses, which obliterate all these chimeras. I dine, I play a game of backgammon, I converse, and am merry with my friends; and when after three or four hours' amusement, I**

would return to these speculations, they appear so cold, and strained, and ridiculous, that I cannot find in my heart to enter into them any farther. (Hume, 1739, 269)

Hume could take refuge in common sense, when he was not doing philosophy, and he found it comfortable to talk like everybody else. He felt uncertain whether his philosophy was doing him or anyone else any good, and knew at least that if he fell in with commonplace thought that he would have a pleasant time of it. Here again there is the choice between the philosophical ascetic life and the social world, but Hume saw the need to move between the two realms. He remained true to his singular sceptical vision in his philosophical writings, but kept in touch with the social world at other times, as he was determined to do, and this is certainly a formula for hanging on to mental health. So, here is Hume long ago, in the middle of the eighteenth century, deliberately adopting a philosophical *persona*, and no less deliberately setting it aside when he is better served by another. What *I* want to say, though it is not quite what Hume says, is that his sense of self is political. It seems actually to be generated through his contact with others. When he isolates himself his sense of self becomes problematic, but when he is engaged with others then his sense of self returns and he is comfortable again. Identity is relational. The procedure is taken further when Hume's writing becomes populated with various *personae* who utter Hume's words in his dialogues and in episodes such as a visit from a 'friend' with principles that 'Hume' – the narrator-character – cannot endorse but which he finds surprisingly persuasive, so one imagines that Hume's views are closer to those of this 'friend' than they are to those of 'Hume' (1751, section XI). When Deleuze and Guattari did this two hundred years later it was seen as postmodern ('Since each of us was several, there was already quite a crowd') but with Hume we can see that it is very traditional. His dialogues are modelled on Cicero's, where again it is unclear which views among the characters might have coincided with Cicero's own. In philosophy the dialogue form goes right back to Plato, where the various views are distributed amongst the monomaniacs who participate in conversation with Socrates. But unless we suppose that these conversations really took place, and that Plato was just the minutes secretary at the Academy, they are acts of ventriloquism on Plato's part, as he adopts different voices in order to get his point across – fictional reworkings and conflations of remarks and arguments that might well have happened, as truthful and as artful as dramatic reconstructions can be.

Deterritorialization

To go back to the example of the hefted sheep: it is as though Hume in adopting his sociable *persona*, which came easily to him and felt 'natural', was comfortably hefted in Edinburgh society when he participated in chit-chat and backgammon. He was bonded to the territory, and knew how to behave as one of the flock, and while he did that he felt good about it. However he could also switch out of it, and his philosophical thought was an act of deterritorialization, taking him out of Edinburgh, out of the world of bonhomie, and into a world he could barely describe to his neighbours. The adoption of various other *personae* were various reterritorializations, as these characters were grounded in ways of thinking that belonged somewhere or other. It is as if our sheep, having by chance or through temporary madness wandered away from the flock, had been able to fall in with a new flock and become hefted in the new place, some lunar landscape. The ascetic condition, which overtakes its victims with mania, alcoholism or anorexia when it becomes too demanding a habit, is (or can be) an act of deterritorialization. Hume, reporting back from his adventures in deterritorialization, tells us what it feels like to be a lost sheep, but once he has returned he is relieved to be safe home again, and in that state of mind he is unsure whether the risks were worth it. That is his comfortable sociable way of speaking like everybody else; but he kept going back. There was another part of him that was drawn back to the exhilaration of finding himself deterritorialized, when anything might happen, and he found himself saying on the page things that to this day retain a compelling lucidity but which he could never have said aloud in the company he enjoyed. Some of his texts were published only after his death, but they are read now, still producing lines of flight in their readers, while the comforts of his backgammon are evoked only on the page. The uncertain grasp on the 'self' in the deterritorialized Hume (the philosophical Hume) contrasts with the unselfconscious confidence with which he moved in society. And in the commonplace world of common sense, he could present the philosophical mood as a mental illness for which a lack of concentration and a willingness to be distracted are the natural cure:

I am first affrighted and confounded with that forelorn solitude, in which I am placed in my philosophy, and fancy myself some strange uncouth monster, who not being able to mingle and unite in society, has been expelled all human commerce, and left utterly abandoned and disconsolate. Fain would I run into the crowd for shelter and warmth; but cannot prevail with myself to mix with such deformity. I call upon others to join me, in order to make a company apart; but no one will hearken to me. Every one keeps at a distance, and dreads that storm, which beats upon me from every side. I have exposed myself to the enmity of all metaphysicians, logicians, mathematicians, and even theologians; and can I wonder at the insults I must suffer? I have declared my disapprobation of their systems; and can I be surprized, if they should express a hatred of mine and of my person? When I look abroad, I foresee on every side, dispute, contradiction, anger, calumny and detraction. When I turn my eye inward, I find nothing but doubt and ignorance. All the world conspires to oppose and contradict me; though such is my weakness, that I feel all my opinions loosen and fall of themselves, when unsupported by the approbation of others. Every step I take is with hesitation, and every new reflection makes me dread an error and absurdity in my reasoning. (Hume, 1739, 264–5)

. . . perhaps one day, this century will be known as Deleuzian.

Now that Hume is routinely taken to be Britain's greatest philosopher, it is touching to read about these doubts, which it was hardly prudent to publish at the time – he was then a young man with the hope of an academic career that never quite materialized. Deleuze and Guattari have in a similar way said outrageous but closely interconnected things, and have stuck to their guns. At times like this one can see the value to them of a statement from Deleuze's friend Michel Foucault, 'perhaps one day, this century will be known as Deleuzian' (Foucault, 1970, 165). However inaccurate, facetious, ironic or jocular it might have been, it was a statement of support that came when it mattered. It is often repeated, and is used – perhaps mainly in blurbs by publishers trying to shift books – without obvious acknowledgement of its absurdity. Deleuze is on the side of the demons, rather than the would-be gods, the transgressors rather than the law-makers, the nomadic war machine rather than the state apparatus. His thought jumps about, and if it were to be codified and made programmatic – something he seems, despite himself, to slip into on occasion – then it loses the mischief and vitality that makes it so appealing. For a long time he studied history of philosophy, he said:

> But I compensated in various ways: by concentrating, in the first place, on authors who challenged the rationalist tradition in this history (and I see a secret link between Lucretius, Hume, Spinoza, and Nietzsche, constituted by their critique of negativity, their cultivation of joy, the hatred of interiority, the externality of forces and relations, the denunciation of power . . . and so on). What I most detested was Hegelianism and dialectics. My book on Kant's different; I like it, I did it as a book about an enemy that tries to show how his system works, its various cogs – the tribunal of Reason, the legitimate exercise of the faculties (our subjection to these being made all the more hypocritical by our being characterized as legislators). But I suppose the main way I coped with it at the time was to see the history of philosophy as a sort of buggery or (it comes to the same thing) immaculate conception. I saw myself as taking an author from behind and giving him a child that would be his own offspring, yet monstrous. It was really important for it to be his own child, because the author had actually to say all I had him saying. But the child was bound to be monstrous too, because it resulted from all sorts of shifting, slipping, dislocations, and hidden emissions that I really enjoyed. I think the book on Bergson's a good example. And there are people these days who laugh at me simply for having written about Bergson at all. It simply shows they don't know enough history. They've no idea how much hatred Bergson managed to stir up in the French university system at the outset and how he became a focus for all sorts of crazy and unconventional people right across the social spectrum. And it's irrelevant whether that's what he actually intended. (Deleuze, 1990, 6)

So Bergson is unexpectedly recruited to a counter-canon of joyous life-affirming transgressors. There is even something to be found in the strait-laced Kant, but this is a very short book, and Deleuze's subtitle for it, '*The Doctrine of the Faculties*', makes use of two loaded words, which mean not only the teaching set out by Kant on his idea of 'faculties', but could also be translated as 'the dogma of the universities', and this play on words hints at the immaculate conception Kant is going to get. Another irony is that Deleuze was employed by universities. He settled at Paris VIII (Vincennes) which is an anomaly in the French educational system – an ultra-liberal establishment (which styled itself a counter-university) that was set up in response to the famous student riots '*les évènements*' of May 1968, in which Guattari had been prominently involved. Guattari sought out Deleuze in 1968, and they started the collaboration that would become *Anti-Oedipus* (1972). Guattari's methods were less sly than Deleuze's. He was prosecuted for 'an outrage to public morals' in 1973, when he published a special issue of a journal, *Recherches*, entitled 'Three Billion Perverts: *Grand Encyclopedia of Homosexualities*' (see Genosko, 2006). His professional life as a psychiatrist was no less contentious. He worked at an experimental clinic which he co-founded with Jean Oury in 1953, at the then ruined chateau of La Borde (Cour-Cheverny, Loire-et-Cher, 100km south of Paris) where they sought to empower the patients, rather than sedate them into passivity. When Deleuze and Guattari worked together, they found themselves doing things that neither would have done separately, despite their shared attitudes.

> **As for the technical side of writing the book together, being two was not a problem for us, it served a precise function, as we came to realize. One thing about books of psychiatry or even psychoanalysis is rather shocking, and that is the pervasive duality between what an alleged mental patient says and what the doctor reports – between the 'case' and the commentary on the case . . . Now we didn't think for a minute of writing a madman's book, but we did write a book in which you no longer know, or need to know, who is speaking: whether it's a doctor, a patient, or some present, past or future madman speaking . . . Strangely enough we tried to get beyond this traditional duality because there were two of us writing. Neither of us was the madman, and neither the doctor: there had to be two of us if we were to uncover a process that would not be reducible to the psychiatrist and his mental patient, or to the mental patient and his psychiatrist. (Deleuze, 2004, 218–19)[3]**

Guattari was voluble, but needed a collaborator if he was going to get his thoughts organized on the page. Deleuze's careful logical working through of philosophical concepts was complementary to Guattari's skills, but the two of them shared a general orientation in favour of the promotion of freedom, life and joy, which they found being stifled by bureaucracies, governments and the ways of thinking that they inculcate in the hefted citizenry. For example, Guattari saw 'an unconscious complicity, an internalization of repression that works in successive stages, from Power to the bureaucrats, from the bureaucrats to the militants, and from the militants to the masses themselves . . . This is what we witnessed after May '68' (in Deleuze, 2002, 217). The point exactly replicates something that Deleuze found in Spinoza. What he found in Guattari was a restless energy:

> **Félix has always possessed multiple dimensions; he participates in many different activities, both psychiatric and political; he does a lot of group work. He is an 'intersection' of groups, like a star. Or perhaps I should compare him to the sea: he always seems to be in motion, sparkling with light. He can jump from one activity to another. He doesn't sleep much, he travels, he never stops. I am more like a hill: I don't move much, I can't manage two projects at once, I obsess over my ideas, and the few movements I do have are internal. I like to write alone, and I don't like to talk much, except during my seminars, when talking serves another purpose. Together, Félix and I would have made a good Sumo wrestler. (Deleuze, 2003, 237)**

Who were Deleuze and Guattari? In a manner of speaking they were a Sumo wrestler. They were a sparkling sea and a slow-moving hill. They were a tendency to dissolve the firm outlines of common sense into something more nebulous, instinct with possibilities. They were a joyous affirmation of the desires that we find we have when we take away the oppressions that have formed us. They were a dissolution of received identities and preconceptions, longitudes and latitudes: speeds, intensities and affects. And, having reached that point, with a wry smile and a twinkle in the eye, they were willing on occasion to adopt the forms of common sense so as to be good company. It's nice, after all, to speak like everyone else.

CHAPTER 2

Machines

Swarming

Machines of every description swarm across the opening pages of *Anti-Oedipus*. Machines of more descriptions than one has previously encountered, assembling themselves, plugging in and turning on, connecting and disconnecting, heating, breathing, lactating:

> **The mouth of the anorexic wavers between several functions: its possessor is uncertain as to whether it is an eating-machine, an anal machine, a talking-machine, or a breathing-machine (asthma attacks). Hence we are all handymen: each with his little machines. For every organ-machine, an energy machine: all the time, flows and interruptions. Judge Schreber has sunbeams in his ass. A *solar anus*. And rest assured that it works: Judge Schreber feels something, produces something, and is capable of explaining the process theoretically. Something is produced: the effects of a machine, not mere metaphors. (Deleuze and Guattari, 1972, 1–2)**

The text has a certain notoriety. 'Félix thinks our book is addressed to people who are now between the ages of seven and fifteen.' said Deleuze, 'Ideally so, because the fact is the book is still too difficult, too cultivated, and makes too many compromises. We weren't able to make it clearer and more direct. However, I'll just point out that the first chapter, which many favourable readers have said is too difficult, does not require any prior knowledge' (Deleuze, 2002, 220). In saying this, Guattari rightly points out that the book's reception requires that it reaches people at an impressionable age. It absolutely requires that readers' established common-sense patterns of thought are disrupted, and if we are to find that stimulating rather than merely annoying then we need to be receptive to the idea that the book is opening up a world that we are eager to know about, and not simply be appalled by it, which is more likely to be characteristic of a teenager than an old buffer. Nevertheless, as is evident in the passage quoted above, it is clear that in addition to the text needing 'parental advisory' warnings, it is dense with allusions. The *solar anus* is a reference to a surreal declamatory text by Georges Bataille. It is a glancing allusion, and even after reading Bataille's text it is not too clear why the allusion would have been made. Perhaps the text was 'in the air' at the time, or perhaps because Deleuze and Guattari had been reading Bataille – as they had, because more serious use

. . . that the spinning of the earth is what drives the pistons of sexual coupling.

is later made of *The Accursed Share* (1949). However in the context of *Anti-Oedipus* one sees the sun as a body and as a machine. One of the images in Bataille's text that resonates strongly with *Anti-Oedipus*, is the sexualized image of the pistons of a steam engine: 'The two primary motions are rotation and sexual movement, whose combination is expressed by the locomotive's wheels and pistons. These two motions are reciprocally transformed, the one into the other' and it seems as if the multiplicity of sex acts on earth are what keeps it turning, or conversely that the spinning of the earth is what drives the pistons of sexual coupling (Bataille, 1931). Volcanoes act as anuses on earth, while the sun, if we see it as a machine, produces sunlight as a human body produces excrement. That idea is spelt out neither in Deleuze and Guattari's text, nor in Bataille's, but it takes shape in me as I think about the texts. The book acts as a machine when I link up with it, producing in me ideas that I would not have had by myself.

The case of Schreber

Schreber, a German judge, published an autobiographical account of his persecution in a futile attempt to persuade others of his sanity (Schreber, 1903). Briefly:

> **Schreber was convinced that God had chosen him to bring the new messiah into the world, and that his body was therefore being turned into that of a woman. He also believed that everyone he met had died, that what he saw**

and heard was only fleeting images of people, ghostly apparitions sent by God to taunt and tempt. When he could no longer bear the ceaseless teasing of aggressive voices, he screamed as loud as he could, which annoyed the neighbours, especially at night, and meant that he had to be restrained.

The case is clear: he suffered from acute paranoia, the main symptom of which is delirium. [. . . But] these symptoms abated when, without in any way renouncing his delirious convictions, he started writing his memoirs, in order to convince those who had committed him to an institution of his sanity, and to present his convictions to a wider audience. (Lecercle, 1985, 1)

Given a certain effect, what machine is capable of producing it? And given a certain machine, what can it be used for?

Schreber is better known than one might think from his plight, because Freud took an interest in his text and presented an analysis of it in one of his 'case histories' (Freud, 1911). Where does this leave us with Deleuze and Guattari's text? One point that can be made is that the allusions in the text certainly repay exploration as they lead the reader on to the most extraordinary material. If I am an 'ideal' impressionable teenage reader, then I will not have come across these texts before, and will be thrilled to find them when I do. The main point being made in the opening pages, though, is to look at the body (integrally with any mind that is involved in it) as a machine or swarm of machines, then to move on, to ask 'Given a certain effect, what machine is capable of producing it? And given a certain machine, what can it be used for?' (Deleuze and Guattari, 1972, 3). In this way the analysis that they propose is entirely pragmatic. Does it work? What can we do with it? They are an architect's most habitual questions, and they seem straightforward enough. What makes the vision complex is the sheer number of things that are drawn into it. We can ask the questions of a body, or a life. We can ask them of an urban space or a view of hills. Remember Deleuze and Guattari's definition of a body: it can be an animal, a body of sounds, a mind or an idea; it can be a linguistic corpus, a social body, a collectivity. We can ask the questions of a plank of wood, or steel girder. The product of one machine is the raw material for another. If we ask 'Does it work? What can we do with it?' of a plank then we get one set of answers, but if we ask these questions of a tree then we do not necessarily find ourselves starting with the plank. (But if we ask the questions of the supplies in a builders' merchant's stock, then it is self-evidently a good idea to go with the plank.) Everything is connected to everything else (as Spinoza taught) and the edges of an identity are formed through interaction and convention. They establish a frame, which is constructed, whether or not we notice the process. They are not natural and self-evident (as we saw in Hume) but if we are deeply immersed in common sense then we might think that they are, just as an unusually self-aware hefted sheep might think that it is free. 'What the schizophrenic experiences . . . [is] nature as a process of production,' say Deleuze and Guattari.

> **It is probable that at a certain level nature and industry are two separate and distinct things: from one point of view, industry is the opposite of nature; from another, industry extracts its raw materials from nature; from yet another, it returns its refuse to nature; and so on. Even within society, this characteristic man-nature, industry-nature, society-nature relationship is responsible for the distinction of relatively autonomous spheres that are called production, distribution, consumption. But in general this entire level of distinctions, examined from the point of view of its formal structures, presupposes (as Marx has demonstrated) not only the existence of capital and the division of labour, but also the false consciousness that the capitalist being necessarily acquires, both of itself and of the supposedly fixed elements within an overall process. For the real truth of the matter – the glaring, sober truth that resides in delirium – is that there is no such thing as relatively independent spheres or circuits. (Deleuze and Guattari, 1972, 3–4)**

What the schizophrenic experiences . . . [is] nature as a process of production.

It was Adam Smith, back in the eighteenth century, who drew attention to the way in which the division of labour into specialized tasks made for vast increases in productivity. By the time that Karl Marx was writing in the nineteenth century this lesson had been very thoroughly implemented in industrial manufacture, so that many of the operations in making a product could be done by machines (in the ordinary sense of the word – Deleuze and Guattari call them 'technical-machines') and they could be tended by people with no overview of their place in the whole process. It is easily possible to perform a routine task without feeling responsible for the far-flung repercussions it might have. (One has to earn a living. If I didn't do it, then someone else would. How was I to know?)

The machine as a whole might be producing desert where there was forest, but each individual in the machine might be doing very well out of it along the way, and would feel that he or she was responding to the abstract logic of the situation.

One of the lessons that Deleuze and Guattari learnt from Marx was to see production as something that happens across and between individuals. It is possible when this happens that no one feels the need to take responsibility for the further consequences. If I am an architect who is just doing what the client wants, then am I, or is she responsible? And if the client only wants what the market wants, then surely she is just being responsible to her shareholders (the concern with 'the bottom line' – large profit or small – profit or loss: *of course* we want to make a profit). And the shareholders are not to be characterized too simply as avaricious capitalists, because these days the important shareholders are pension funds and suchlike, who want to see a good return on their investments so that the elderly people who are dependent upon them can have a decent standard of living; so if the business is not being managed profitably they will take their money away and make their investment elsewhere. Everyone can be taking rational decisions, and yet the result could be the most rapacious commercial exploitation of a place. The machine as a whole might be producing desert where there was forest, but each individual in the machine might be doing very well out of it along the way, and would feel that he or she was responding to the abstract logic of the situation. We might expect a higher authority, like a government, to take action, but if the decision were unpopular

Once the machines are assembled, they have an identity and a life of their own.

with all the people who were doing well, and the people who depended on them, then the government might be removed. Once the machines are assembled, they have an identity and a life of their own.

The book of the machines

This is an idea that was explored in a brilliant and clear-headed exposition of the view of a machine as an organism: 'A profound text by Samuel Butler, "The Book of the Machines"' (Deleuze and Guattari, 1972, 284). Of all the texts referenced and quoted in *Anti-Oedipus*, this is the one that is directly quoted at greatest length (264–5) and it is worth reading in full.[1] Butler's point is that machines have lives of their own, and the fact they are not organic is irrelevant to that insight. 'Does anyone say that the red clover has no reproductive system because the bumblebee (and the bumblebee only) must aid and abet it before it can reproduce? No one. Each one of us has sprung from minute animalcules whose entity was entirely different from our own . . . These creatures are part of our reproductive system; then why are we not part of that of the machines?' (Butler, in Deleuze and Guattari, 1972, 285).[2]

The relation that Deleuze and Guattari found compelling and which they used repeatedly was that between the wasp and the orchid – more precisely the relation between the wasp-orchid and the Thynnine wasp. Part of the orchid has evolved so as closely to resemble the female of the wasp species, and when the male wasp tries to mate with it he picks up pollen which he transfers to the next wasp-orchid that seduces him. This is beautifully shown in the film *Adaptation* (2002) where the wasps and orchids can be as difficult to tell apart as the film's protagonist and his twin brother (see below).[3] Whenever Deleuze and Guattari refer to the wasp and the orchid, they always seem to suppose that we have heard about this pairing before; their point, though, is Butler's point: that the two have developed inseparably – the example is intensified because the plant's adaptation of its form makes the interdependence very apparent. The wasp is evidently and obviously part of the plant, so where should we draw the frame around its 'identity'? Deleuze and Guattari would resist drawing the line. The two connect and are part of the process of production (reproduction) of the machine. In the same way Butler's 'vapour engines' – the advanced technology of his day – depended for their reproduction and evolution on human agents. They depended not only on engineers to tend, and design, and improve the machines, but also on humans burrowing underground to find the fuel that the machines needed, and to smelt

The machine is composed of organic and inorganic parts, which act together to constitute its life and to produce its power and speed.

the iron for the rails that are inseparably a part of this machine. The machine is composed of organic and inorganic parts, which act together to constitute its life and to produce its power and speed. Now that we have computers to help

A Thynnine wasp (*neozeleboria cryptoides*) and a wasp-orchid (*chiloglottis trilabra*) in *Adaptation*, directed by Spike Jonze, 2002.

us think, we seem more than ever to be dependent upon and implicated in the life of machines. There are times when they can seem to run our lives, as they exact demands from us that we would not otherwise have thought of asking from ourselves. The pattern of shift working was invented by Richard Arkwright in the 1770s, as a way of keeping his expensive cotton-machines continuously productive (Fitton, 1989). This seems to mark a shift across a threshold in the balance of power between men and machines, where the machines' demands seem to outweigh the humans'. One is reminded of Ruskin's drowned man: 'lately in a wreck of a Californian ship, one of the passengers fastened a belt about him with two hundred pounds of gold in it, with which he was found afterwards at the bottom. Now, as he was sinking – had he the gold? or had the gold him?' (Ruskin, 1862, 210). Do the machines work for us, or do we work

Do the machines work for us, or do we work for them? How would we tell if it were the other way round? And would the machine agree?

for them? How would we tell if it were the other way round? And would the machine agree? Shift work is now ingrained as a part of life, and there are various practical websites explaining how to cope with it. One way to cope now is to distribute the work around the world, to different time zones, so that a task to be worked out can be considered on the other side of the world while I am relaxing and asleep, and I can pick up the information I need when I start again in the morning. The connections between things become paramount – between the engine and the engineer, between the wasp and the orchid, between the workers in the distributed office network. Now that we routinely make connections electronically through networks, the point does not have to be laboured; but their importance must not be underestimated. It is through the connections that new identities and bodies are formed.

Down with trees

The particular network structure that Deleuze and Guattari promote is the rhizome – a plant structure that can bifurcate and send out a new shoot at any point. They use it as a contrast with the tree structure, where everything branches out from a central trunk – the little twigs branch out from larger ones, and so on, back to the sturdy core. This is treated as an image of centralized power, or as something more than an image: it is a model of centralization, a real acting-out of it. Trees are normally given a good press by the environmentally concerned, and it is surprising to find them so roundly condemned.

. . . in the West, the tree has implanted itself in our bodies.

> **It is odd how the tree has dominated Western reality and all of Western thought, from botany to anatomy, but also gnosiology, theology, all of philosophy . . . : the root-foundation, *Grund, racine, fondement.* The West has a special relation to the forest, and deforestation; the fields carved from the forest are populated with seed plants produced by cultivation based on species lineages of the arborescent type; animal raising, carried out on fallow fields, selects lineages forming an entire animal arborescence. The East presents a different figure: a relation to the steppe and the garden (or in some cases the desert and the oasis), rather than forest and field; cultivation of tubers by fragmentation of the individual; a casting aside or bracketing of animal raising, which is confined to closed spaces or pushed out to the steppes of the nomads. . . . Does not the East, Oceania in particular, offer something like a rhizomatic model opposed in every respect to the Western model of the tree? André Haudricourt even sees this as the basis for the opposition between the moralities or philosophies of transcendence dear to the West and the immanent ones of the East: the God who sows and reaps, as opposed to the God who replants and unearths (replanting offshoots versus sowing of seeds).[4] Transcendence: a specifically European disease. Neither is music the same, the music of the earth is different, as is sexuality: seed plants, even those with two sexes in the same plant, subjugate sexuality to the reproductive model; the rhizome, on the other hand, is a liberation of sexuality not only from reproduction but also from genitality. Here in the West, the tree has implanted itself in our bodies, rigidifying and stratifying even the sexes. We have lost the rhizome, or the grass. (Deleuze and Guattari, 1980, 18)**

So these habits of thought, once they are planted in us, take over and refract our view of the world and all our dealings with it. It is probably becoming clear by now how the 'capitalism and schizophrenia' project, across the two volumes *Anti-Oedipus* and *A Thousand Plateaus*, was caught up in every aspect of life. It is set up not as a set of dogmas or even of questions, but as a set of values. It is a work of ethics, and the link with Spinoza's *Ethics* is strong. It too is a project based around immanence rather than transcendence. The 'desiring-machines' that figure so prominently at the opening of *Anti-Oedipus*, are the machines that operate without our noticing them to produce the desires that we do notice, and that we would like to act upon. But as mechanisms that operate to produce consciousness, the machines can be pulling in different directions and producing incompatible desires, which might be resolved at a preconscious level, or might surface as conflicted conscious desires. There are thousands

It is set up not as a set of dogmas or even of questions, but as a set of values. It is a work of ethics.

upon thousands of these mechanisms, of which we become aware only as they produce effects that approach the level of consciousness, and what goes on amongst them is a micropolitics – thousands upon thousands of rhizomatic connections, without any clear limit on where the connections would stop, and without any necessity to pass through a centralized arborescent hub. The scale of operations builds up from a preconscious sub-'individual', who is already a swarm of desiring-machines, to a social group, or a crowd, where certain aspects of the people involved connect together to produce a crowd-identity that is unlike that of any of the individuals in the crowd. Crowds will do things that individual people would not (Canetti, 1973). The individuals are to the crowd what desiring-machines are to the individual. Except that one could alternatively say that it is certain of the mass of individuals' desiring-machines that, upon being brought together in the crowd, they are found to be able to act together to produce the group identity. The crowd is a body. Some of the mechanisms that would come into play in the individual acting alone are somehow switched out of the circuit, and become irrelevant to the crowd, and having been switched off cannot inhibit the crowd's actions. So the sense of the 'individual' is even further problematized, and we see it to be highly divisible. But nevertheless the idea of the individual is deeply ingrained in our language, and if we're trying to explain ourselves, we might find that it's the most direct word to be using. If we're trying to connect with others then we need to be able to allow ourselves, from time to time, to speak like everyone else. As we follow Deleuze and Guattari further into their world, it becomes increasingly difficult to do that, as each 'straightforward' utterance seems, from an alternate view, to have an inaccurate aspect.

Abstract machine

Deleuze and Guattari's ways of thinking are above all practical ('Given a certain machine, what can it be used for?' 1972, 3) but they aim for a high level of generality. If I recognize the same pattern in the way that my psychology builds up from a swarm of interconnected desiring-machines, and in the way that a crowd-psychology can form in an assembly of individuals, then I can take the view that the same mechanism is at work in both cases. The important thing here is the mechanism – the machine – that brings about these effects. It is a real thing that is at work, producing these effects, but it is abstract. It is embodied in the different people and particular crowds that one has encountered or heard about. We might notice it in ourselves when we act impulsively and unreasonably, for example by making romantic attachments, or by being unable to concentrate on a boring but necessary task. From a calmly rational point of view, Dido, the queen of ancient Carthage, acted unwisely when she committed suicide after Aeneas left her, and might have been saved by counselling (though her romantic reputation would have perished). The wild girl of Châlons did not mean to eat her sister, but couldn't help herself (Ballantyne, 2005, 112). Crowds can

The wild girl of Châlons did not mean to eat her sister, but couldn't help herself.

run out of control in ways that individuals rarely do, whether it be a football crowd, a Nazi rally or the Bacchantes who ran wild and tore Orpheus limb from limb. The examples multiply, but the mechanism is broadly the same – a politics operates to make connections between the various parts, and it is a dispersed politics, so the unstoppable will of the crowd, or my uncontrollable urge, is acted upon despite some other part having a different inclination. Deleuze and Guattari especially resist the idea that the abstract machine exists in some way prior to its embodiment in particular cases, as Plato had argued with his theory of Forms (Deleuze and Guattari, 1980, 510). The abstract machine is always embodied in the example, one example after another.

Immanence

The polarity, immanence versus transcendence, which is mentioned above, emerges as an important theme in Deleuze and Guattari's work. Immanent properties inhere in things, and are there all along, though it might take special circumstances to make them apparent. Transcendent properties come from outside, usually from a divinity or the spirit world. One summer I was in a field after harvest, when the weather was unusually hot. The field was covered in wheat stubble and straw, and the dusty dried leaves that had crumbled away. The air was still and hot. There were quite a few other people around, because there was a fair in the field at the time. There was a horse-powered threshing machine, and a display of country dancing; but it was so debilitatingly hot that it all seemed a bit half-hearted, because

The dusty leaves seemed to be moving into formation. They made a long fairly straight line across the ground, lifting at one end.

everybody knew that they would really rather be sitting in the shade with a cold drink. The sun glared relentlessly, and the surface of the ground was hottest of all. Something caught my eye. The dusty leaves seemed to be moving into formation. They made a long fairly straight line across the ground, lifting at one end. It looked a bit like a rope, frayed at the rising end. It started moving across the field, and turned to a more upright position, having become recognizable now as a small tornado. It maintained its form, a vertical pillar of fragments of dessicated vegetation, perhaps 6 metres tall. It reached the edge of the field, vigorously agitated the leaves of the trees there, and then dispersed – just disappeared without trace – and the leaves on the trees became calm again; and the sun beat down; and nobody else who was there noticed that anything had happened.

It is quite possible to see this little tornado as a result of the conditions that were in place at the time. I infer, from the way things behaved, that there must have been convection currents in the air, creating a strong updraught that started twisting round for exactly the same reasons that water twists round when it goes down a plug hole. Maybe these things happen all the time, and no one pays any attention. That would be an explanation that the phenomenon was immanent in the conditions. On the other hand, if I had watched Andrei Tarkovsky's film *Mirror* (1975) I would have seen a scene in which a still field with long grass, bounded by trees, is suddenly and inexplicably overtaken by wind, which agitates the scene violently, and then moves on. There is no commentary, we are not told what to think about the scene. But somehow one comes away with the sense that the scene has spiritual significance – that there has been some sort of visitation – which is a transcendental explanation; it is a fine example of 'transcendental style' in film (Loughlin, 2003).

. . . somehow one comes away with the sense that the scene has spiritual significance – that there has been some sort of visitation.

Somehow it was not the explanation I reached for when I encountered the little whirlwind. This might be because, despite the naturalistic appearances, the occurrence in *Mirror* is so unexpected and so powerful as to seem supernatural. It was achieved – one finds by paying attention to supporting material on the DVD, and noticing the shadow of a helicopter that figures nowhere in the film's images – by means of a helicopter's downdraught. The helicopter itself remained out of shot, and the noise it made is substituted on the soundtrack by a recording of natural wind. So indeed there is a transcendental operation in play – the hand of the film-maker – acting as *deus ex machina*. In nature there are surprises, but there are no 'special effects'.

Spinoza was excommunicated by the Jewish community in Amsterdam because he said that God was immanent in nature. He is therefore strongly associated with the idea of immanence, and this is why, from time to time, Deleuze and Guattari liked to call themselves 'Spinozists'. The sciences discuss immanence when they analyse 'emergence'. Emergent properties develop in a system when it is allowed to run, and they are interesting when they are unexpected properties that do not seem to have been programmed into the start conditions. Immanence and emergence are therefore different aspects of the same process, relating a set of generative properties to a set of products. For example, a swarm of slime mould, an amoeba-like organism with no discernible brain, can find the shortest route through a maze, despite the fact that an individual amoeba would not be able to do it. Acting as a group, somehow it becomes possible for them to do it, and not because of any supernatural or psychic guidance that the slime mould can tap into. Its actions can be modelled electronically, and the beauty of the electronic model is that one can know exactly what generative properties have been given. All they have to do is react with their very limited sensory apparatus to the stimuli that affect them, especially the stimuli from the other slime-mould organisms around them.

. . . a swarm of slime mould, an amoeba-like organism with no discernible brain, can find the shortest route through a maze, despite the fact that an individual amoeba would not be able to do it.

Individually they cannot do anything like 'think', but collectively they can (Johnson, 2001, 11–17). The brain can be seen to work in a way that has its parallels with this, as is evident from the title of a book by Marvin Minsky: *The Society of Mind* (1985). Minsky's work is in artificial intelligence, and in

We are political all the way down to the unconscious bodily responses that we could not call 'thoughts'.

this book he presents a series of surprisingly simple decision-making or 'recognition' moves that when connected together in vast numbers – something of which the brain is certainly capable – do something like 'thinking'. A more recent book, *The Emotion Machine* (Minsky, 2006) takes the thinking further. The human being is an emotion machine, he says, and maybe computers could be. Far from being 'individual' our minds are already multiplicities that work together as some sort of society. We are political all the way down to the unconscious bodily responses that we could not call 'thoughts', and to the things (which might be outside our bodies) that prompt those responses.

Network

In giving such prominence to the idea of a vast network of interconnectivity when they did, in the early 1970s,[5] Deleuze and Guattari were ahead of the explosion of interest that there would be as the internet grew in cultural significance, particularly – at a popular level – from the late 1980s. The idea has moved from seeming very arcane, to being a very evident part of everyday life, as a certain group of our connections in the world become much more evident to us, and we construct our identities in part through those connections. At the scale of international politics, the exhortations against centralized arborescent structures are well taken, and the effects of electronic networks are generally experienced as empowering and helpful at a local level. What distinguishes Deleuze and Guattari's account is the fact that it connects these large-scale networks with the networks in the body, and between bodies, so that we start to see things like temperament and identity as emergent properties that are products of the machines immanent in the initial conditions. Matter, environment and emergent consciousness all connect across to something nearby, and ultimately to everything else, to give a political and ethical dimension to the thinking. The thought in *Capitalism and Schizophrenia* seems at times to be scientifically informed, and at other times to be crazily associative, clutching at poetic impressions or making mad conjectures. But more importantly it is conceptual. It is a very sustained and vigorous presentation of concepts (not 'ideas', in case we mistake them for Plato's transcendent Ideas, but definitely 'concepts') that have proved to be fruitful, and which seem to be very far from exhausted.

The body

The view of the body that comes in with Deleuze and Guattari's analysis is equally open-ended. If we look at the body in terms of machines, then – as Samuel Butler pointed out – there is no reason to suppose that there is any point in saying that what a body can do is limited to saying what it can do using only its organic parts. I can dig much better if I pick up a spade. I can see further if I look through a telescope. These prosthetics extend and amplify the body's capabilities. I can put them down and leave them behind, but when I need them they become part of me – part of the digging-machine, or the seeing-machine – and in a way they are always part of me if I use them. Indeed if I have developed the habit of using them and they then become unavailable to me, I will feel their lack, and will feel disabled, in the same way (but less painfully) as if I had lost a hand or an eye.

Renaissance drawings show grids of squares with human figures superimposed across them, which turn into the ground plans of churches, and in doing so embody divinely ordered proportion. The body in the Deleuze-and-Guattari-world is utterly different. It shits and fucks.

It has been one of the most enduring habits of Western architecture to see human form reflected in buildings (Rykwert, 1996). But our sense of what human form is, has not been uniform at all times and in all places. In the high culture of Western architecture the tendency has been to isolate the proportions of an idealized figure, and to suppose that by applying them in designing a building that something fundamental and important has been transferred from the human form to the building form. Vitruvius inscribed a square and a circle around the human figure, and these forms were seen to embody something important about the human form, which, to a superficial view, is neither square nor circular. Renaissance drawings show grids of squares with human figures superimposed across them, which turn into the ground plans of churches, and in doing so embody divinely ordered proportion. The body in the Deleuze-and-Guattari-world is utterly different. It shits and fucks, is engaged in processes of production and consumption, has an interior as well as an exterior (or, rather, the interior and exterior are indistinguishable) and it connects in multifarious ways, within itself and with its surroundings.

In its most elemental state it is the 'body without organs' – a term which Deleuze and Guattari adapt into an abstract concept, reterritorializing it in many contexts – but its origin is in a concrete example. It emerged in the last work by the dramaturge of the Theatre of Cruelty, Antonin Artaud (1895–1948). *To Have Done With the Judgement of God* was a rant against America and God, that carries the scars of Artaud's tormented years in lunatic asylums. It was intended for radio broadcast on November 28, 1947, but was suppressed.[6] '. . . there is nothing more useless than an organ. When you will have made him a body without organs, then you will have delivered him from all his automatic reactions and restored him to his true freedom' (Artaud, 1947, 571). So the body without organs is presented here as an idealized state, in which anything becomes possible. (Under a more common-sense description such a body is comatose and has severe psychiatric disorders.) It is the condition of the lost sheep, away from the flock, deterritorialized and desocialized, without politics or a self. At this moment of confusion, it has lost the habits that have been there as part of its ancestral inheritance and its upbringing. But in Artaud's case it goes further. He remembered having found himself during a mental breakdown, with no shape or form, right there where he was at that moment (Deleuze and Guattari, 1972, 8). He was away from the flock of other socialized people, who structured his sense of who he was in the world, and away from the flock of desiring machines that normally structured his sense of who he was in himself. His identity had gone.

He lived for a long time without a stomach, without intestines, almost without lungs, with a torn oesophagus, without a bladder, and with shattered ribs.

This sense of the body without organs, a catatonic body that is not structured by interactions, responses or concepts, is taken up by Deleuze and Guattari and is itself deterritorialized, so that it becomes a mobile concept, signalling in general the removal of all acquired habits and identity. We make ourselves bodies without organs by flirting with catatonia, by suspending our identity. We step out of the world of the actual, the world of common-sense stability, where we function well by repeating the habits of the day before, into the world of the virtual, where anything can happen. It was what Hume described himself as doing when he was thinking philosophically about himself, and consequently losing his sense of his self. Judge Schreber also had problems conceptualizing his body and what was happening to it. 'He lived for a long time without a stomach, without intestines, almost without lungs, with a torn oesophagus, without a bladder, and with shattered ribs, he used sometimes to swallow parts of his larynx with his food, etc. but divine miracles ("rays") always restored what had been destroyed' (Freud, 1911, 147; quoted by Deleuze and Guattari, 1980, 150). The body without organs is *virtually* all the things we could be, but when we're in that state we're *actually* none of them. In order to actualize a virtuality, we need to conceptualize some step towards it as a possibility, and as a body without organs we have no concepts, so we are trapped in a catatonic state for as long as it lasts. The virtual is the realm of the pre-possible, where there is no conception of what the alternative possibilities could be, so if anything happens, it happens without having those possibilities to guide or inform it. It is the soup from which the emergent properties will in due course emerge, but with no sense as yet of what those emerging properties are going to be. Schreber's non-standard actualization was an oddity that tells us something about the range of possibilities, and different cultures at different times have conceptualized the body in multifarious ways (see e.g. Feher, 1989). The body without organs lies before and beyond all the actual alternatives:

without organs is what remains when you take everything away. take away is precisely the phantasy, and significances and :ations as a whole. Psychoanalysis does the opposite: it translates everything into phantasies, it converts everything into phantasy, it retains the phantasy. It royally botches the real, because it botches the body without organs. (Deleuze and Guattari, 1980, 151)

The body without organs is pure immanence ('the plane of immanence') having in it no conceptual apparatus that has been imposed from outside – nothing transcendental about it. 'After all, isn't Spinoza's *Ethics* the great book of the body without organs?' (Deleuze and Guattari, 1980, 153). 'All bodies without organs pay homage to Spinoza' (154). The body without organs is a state of creativity, where preconceptions are set aside. It is the state before a design takes shape, where all possibilities are immanent, and one holds at bay the common-sense expectations of what the design should be. When a stimulus or an internal pain prompts a line of flight, then formations assemble, giving the beginnings of a form – a structure, a detail, a *leitmotif*. The aim could be that the design would be entirely immanent in its initial conditions, and would emerge as a product of the various forces in play in the *milieu*. It would not be imposed from outside as a specified form, but would work with the grain of its matter, from within, but also seamlessly with the *milieu* and networks extending to its horizons. It can crystallize in various ways, at a molecular level, to aggregate and produce different surface effects when it becomes apparent to the senses in a wider world. It is clearest in the sort of house that is a continuation of the person who lives in it, as a mollusc lives in its shell.

The body without organs is a state of creativity, where preconceptions are set aside. It is the state before a design takes shape, where all possibilities are immanent, and one holds at bay the common-sense expectations of what the design should be.

'A house,' said Eileen Gray, 'is man's shell, his continuation, his spreading out, his spiritual emanation.'[7] Our dwellings make it possible for us to live the lives we lead. In a different place, with different arrangements, it would be a different life, with other connections, other opportunities, other obstacles. And with reference to Butler, we might want to ask who would want to claim that an inhabited house was not a living thing? It is animated by the machines that dwell in it and live through it as surely as the body is animated and structured by its desiring machines. Where do we draw the line between our categories? The logic that is generated by Deleuze and Guattari's redescription makes them dissolve into one another, and the dwellings become entities with their own inclinations and desires, as manifested in their behaviour. The houses we dwell in and those that we visit are emotion machines that are animated by our being implicated in them.

CHAPTER 3

House

Plateau

A 'plateau' is a space where forces interact with one another in a relatively stable way, without interference from outside. Conditions may change, but the changes will be worked out from within. Emergent phenomena are produced, but the system is not given a purpose from beyond itself, nor is it cut off to suit the needs of something outside. There is an idea of perpetual stability, so that one could wander away from it and return and find it much the same as before. It is associated with meditative states of mind, or indeed with catatonia, as a body without organs is a plateau. Deleuze and Guattari developed the idea from Gregory Bateson's analysis of the value-system of Balinese culture, which is presented as profoundly different from Western culture in that it tends to look for ways to maintain a steady state, and keep forces in balance, rather than to maximize one value at the expense of others.[1] So the music is hypnotic, rather than climactic, and wealth is not accumulated, but is joyously consumed. The plateau is also part of the earth, which is a loaded term in the Deleuze and Guattari world (as in the *terre* of territorialization, which would re-form the body without organs into its new configuration).

Alongside this, there is the concept of nomadic thought, which wanders across the 'earth', by making deterritorializations and finding new reterritorializations. Nomadic thought in the Deleuze-and-Guattari-world is not a matter of making long journeys around the world for the sake of travel. On the contrary, it could happen without stepping outside one's apartment. It has more to do with the state of mind of the hefted sheep who has wandered away from the territory, and – for as long as it has lost its bearings – has become nomadic. Most often this sort of deterritorialization is immediately recuperated by a reterritorialization, so that one switches from one common-sense regime to another. One's thought is becoming nomadic in that switch, and is fully nomadic if one has become 'at home' in that state in between, so that it becomes habitual and identity-defining. To be always in between territorializations in this way, though, will have practical drawbacks when one has to deal with other people, so it becomes useful to be able to 'visit' one sort of common sense or another, and to 'speak like everyone else' as occasion demands, before wandering away.

Paul Klee, *The Twittering Machine*, 1922.

Architects do this all the time, in moving between the different worlds of the people who participate in a building's becoming. One group of people – engineers, quantity surveyors and builders – works together to see the building constructed, and each professison has its own characteristic vocabulary, its characteristic attitudes and concepts. These different sets of concepts enable each profession to deploy its own expertise.

Then there is another group of interested parties, the representatives of a wider society – town planners, building control officers, fire inspectors and so on – who ensure that the building is one that will not endanger its users or inconvenience their neighbours. They too have their own ways of thinking and speaking, as do the very different sorts of people who commission buildings and who use them. In a commercial *milieu* the building will not happen at all unless the accountants can see a case for it, and when it is configured in those terms the building sounds like a very different thing from the building as it is encountered by the receptionist, the cleaner or the bureaucrat who ends up working in it. A building can be described and redescribed so as to make sense in various perspectives. Ideally it will make good sense in all of them, but in actuality the results can be uneven. In Deleuze-and-Guattari terminology, these different ways of thinking map the building on different planes. (In the French there is no difference between a 'plan' and a 'plane', but Deleuze and Guattari's translators use 'plane'. Had they used 'plan' then the architectural derivation of the concept would have been more evident.) The 'plan' that one uses for discussing spatial arrangements with the client, is different from the 'plan' that one issues to the electrician to show what facilities are needed where. Or the description that one gives in the plane of building construction is radically different from the description that one gives in the plane of cost. These planes intersect, because if for example I make a change to the way the building is constructed then it will have implications for the building's cost, but to make a change in one plane does not necessarily produce a predictable change in the other.

In Deleuze and Guattari's work there is an expression '*se rabat sur*' meaning literally 'falls back onto' which is an expression in projective geometry. If I take a line of a fixed length that is running through space at an angle that is not parallel to a plane, then I can project the line on to the plane, and it will show up on the plane as a line that is shorter than the original – how much shorter will depend on the angle. This is what happens every time one draws a ramp on a plan, or shows in elevation a wall that runs obliquely to the elevation's plane. So using a literal plan of a building, I might increase the height of the structure by ten storeys, and on plan the change might have little discernible effect, just thickening the structural columns, perhaps, and increasing the number of lifts; meanwhile on another plane – the plane of the elevation – there would be a marked change in the building's profile. Then, to extend the system in the way that Deleuze and Guattari do, I could, say, make a change in the construction of the building's structural frame from concrete to steel and it might 'fall back onto' the plane of cost as a small reduction. Or if I were able to increase the running speed of the lifts, there might be a significant increase on the plane of 'user satisfaction'. *Etcetera*. The movements can be non-linear, as in the movement of a puppeteer's hand, which does not mimic or represent the movement that is made by the marionette on the stage below (Deleuze, 2003, 11). This is all quite straightforward; but it becomes more complex when one finds that the plane to which Deleuze and Guattari continually refer is the 'plane of consistency', which is the plane of deterritorialization, of the body without organs, the plane of 'becoming', where we are outside the common-sense world altogether. Deleuze and Guattari keep gravitating back to this deterritorialized state of unformed virtuality, out of which the actual will emerge when desiring machines act upon it.

Actual buildings

A building is a machine, in the same way that Deleuze and Guattari's book is a machine. When I encounter a building, it produces in me certain affects – lines of flight, deterritorializations, whatever. Precisely what affects it produces in me will depend on what I bring to it as part of me – my experience, ideas that I have picked up from reading, stray images that the building calls to mind. Part of this baggage is personal. Perhaps the building reminds me of a place I knew as a child, where I was happy; or perhaps it evokes a place where I was attacked out of the blue. If it happens to do such things then the building might produce in me powerful affects that are a real part of my response – my pulse rate might quicken, I might hyperventilate – and that might be the overwhelmingly important part of the response so far as I am concerned; but such a response would not have any wider significance. It would have been no part of the designer's intention and others would not share it. (*I* feel it as a real response – it's *my* response – but *you* tell me that I'm just imagining it. And of course, in a manner of speaking, that's just what I am doing. I'm imagining it, but I'm imagining it because of what the building is doing to me, which makes it real enough *for me*.) Other sorts of responses come about in ways that can be anticipated or cultivated. If I have studied architecture and recognize that the building before me makes use of the vocabulary of form developed by, say, Louis Kahn, then I will recognize it as a building of some sophistication and ambition on the part of the architect. I would be able to do this because part of my life experience has been the deliberate acquisition of a certain familiarity with these forms. Had I not spent some time doing that, then I would encounter the building differently, perhaps recognizing something special about the building, perhaps not. The point is that the affects are produced, and they are real, but they are not produced by the building acting alone. They are produced when the building and the person come into contact, and people are 'prepared' in different ways by their life experiences, including their education (the French word is more evocative: their '*formation*', which could be translated as 'training'). A building, like any work of art, is a bloc of sensations and affects. An encounter is an experience, an experiment. The two English words 'experience', 'experiment' are one '*expérience*' in French (*avoir une expérience* – to have an experience, *faire une expérience* – to make an experiment) so in the Deleuze-and-Guattari-world, life's experiences are also always experiments. How one reacts to the building will depend upon what one thinks one is engaging with – what 'architecture' we infer in the building. Take a very simple building: a small dwelling built by a poor woodman. It is built using the materials that are at hand – stones taken from the fields, timber from the forest – and is made soundly but with nothing extraneous about it. It is evoked by John Clare (1793–1864) as a place of frugal contentment. At the end of the laborious working day he retires 'to seek his corner chair and warm snug cottage fire' ('*The Woodman*', 1819, line 135) and would rather be there with his children than be crowned king of England. His wife is an important part of the picture, and makes the dwelling work, as he described using his non-standard spelling:

> Soon suppers down the thrifty wife seeks out
> Her little jobs of family concerns
> Chiding her children rabbling about
> Says they'll 'stroy more then what their father earns
> And their torn clohs she bodges up and darns
> For desent women cannot bear the sight
> Of dirty houses and of ragged bairns
> Tis their employment and their chief delight
> To keep their cots and childern neat and tight. (lines 154–62)

The year of Clare's birth was the year of the execution of Marie Antoinette, Queen of France, who famously commissioned a hamlet of cottages at Versailles, so that she could escape from the rigid formalities of courtly life and convince herself, if no one else, that she could participate in a simple way of life as a milkmaid. Whatever similarities there might have been between the buildings of Clare's woodman's cot and Marie Antoinette's, we can be sure that they connected into different architectures. To go back to the idea of the machine: the 'woodman-cottage machine' produces something very different from the 'Marie Antoinette-cottage' machine. The woodman-cottage is the woodman's shell, part of his basic equipment for survival in the world. The Marie Antoinette-cottage is a *cottage orné*, an ornamental building that may not be a dwelling at all, but which is visited for the sake of its effectiveness in producing sentiment. Indeed, given the role assigned to milkmaids in eighteenth-century literature, their innocence and *naïveté* amounting to a form of coquetry (a role that would later by played by the stereotype of the 'French maid' in British farces) there was certainly something romantic and erotic about Marie Antoinette's hamlet. There is nothing fixed about the architecture in relation to the building. It depends upon which machine it has been assembled into. The woodman-cottage is the centre of a world, where folk tales are told beside the fire. The *cottage orné* is visited for pleasure, or for contemplative isolation, a place for trysts and for books. So if Clare's woodman moved out and Marie Antoinette moved in, it would be used in different ways, a different machine would be constituted, and it would produce different affects (Arnold and Ballantyne, 2004). Clare's poetry was probably read in more *cottages ornés* than woodmen's cottages. In the

Deleuze and Guattari world, art has prehuman origins; it begins with the house, and with song:

> Perhaps art begins with the animal, at least with the animal that carves out a territory and constructs a house (both are correlative, or even one and the same, in what is called a habitat). The territory-house system transforms a number of organic functions – sexuality, procreation, aggression, feeding. But this transformation does not explain the appearance of the territory and the house; rather it is the other way around: the territory implies the emergence of pure sensory qualities, of *sensibilia* that cease to be merely functional and become expressive features, making possible a transformation of functions. No doubt this expressiveness is already diffused in life, and the simple field of lilies might be said to celebrate the glory of the skies. But with the territory and the house it becomes constructive and erects ritual monuments of an animal mass that celebrates qualities before extracting new causalities and finalities from them. This emergence of pure sensory qualities is already art, not only in the treatment of external materials but in the body's postures and colours, in the songs and cries that mark out the territory. It is an outpouring of features, colours and sounds that are inseparable insofar as they become expressive (philosophical concept of territory). Every morning the *Scenopoetes dentirostris*, a bird of the Australian rain forests, cuts leaves, makes them fall to the ground, and turns them over so that the paler, internal side contrasts with the earth. In this way it constructs a stage for itself like a ready-made; and directly above, on a creeper or a branch, while fluffing out the feathers beneath its beak to reveal their yellow roots, it sings a complex song made up from its own notes and, at intervals, those of other birds that it imitates: it is a complete artist. (Deleuze and Guattari, 1994, 183–4)[2]

Deleuze and Guattari discuss the structuring effect of the refrain (*ritournelle*) in *A Thousand Plateaus*:

> A child in the dark, gripped by fear, comforts himself by singing under his breath. He walks and halts to his song. Lost, he takes shelter, or orients

> himself with his little song as best he can. The song is like a rough sketch of a calming and stabilizing, calm and stable, centre in the heart of chaos. Perhaps the child skips as he sings, hastens or slows his pace. But the song itself is already a skip: it jumps from chaos to the beginnings of order in chaos and is in danger of breaking apart at any moment. There is always sonority in Ariadne's thread. Or the song of Orpheus. (Deleuze and Guattari, 1980, 311).

The outer world is kept at bay, or is allowed inside in a filtered and controlled way, allowing in something from outside when it will help, but protecting the activity within:

> Sonorous or vocal components are very important: a wall of sound, or at least a wall with some sonic bricks in it. A child hums to summon the strength for the schoolwork she has to hand in. A housewife sings to herself, or listens to the radio, as she marshals the anti-chaos forces of her work. Radios and television sets are like sound walls around every household and mark territories (the neighbour complains when it gets too loud). For sublime deeds like the foundation of a city or the fabrication of a golem, one draws a circle, or better yet walks in a circle as in a children's dance, combining rhythmic vowels and consonants that correspond to the interior forces of creation as to the differentiated parts of an organism. A mistake in speed, rhythm, or harmony would be catastrophic because it would bring back the forces of chaos, destroying both creator and creation. (Deleuze and Guattari, 1980, 311)

Once this 'house' is established, one can venture out from it and engage with the outside world. One can see the generative and emergent role of the refrain – the little tune – informed by the heartbeat, and the rushing of the blood; but it involves an engagement with the outside also, with an outlook on the universe. Deleuze and Guattari are informed in their thinking by the work of André Leroi-Gourhan (1911–86) a palaeontologist who excavated, and who speculated in a well-informed way about the emergence of early cultural phenomena – and indeed about the emergence of man: some of the speculation is concerned with the prehuman. He gave importance to the phenomena of rhythm in his discussions of emergent conceptions of the world, for example in connection with the earliest maintained dwellings (Leroi-Gourhan, 1964, 314) and he proposed for example two alternate spatial schemata: the concentric and the radial. Concentric space is settled territorial space, that establishes an idea of a centre – a granary – around which circles are inscribed, and 'that enables us, while remaining immobile, to reconstitute circles around ourselves extending to the limits of the unknown' (325–6). Conversely, radial space is the space of the nomadic hunter-gatherer, who moves across the surface of a territory, offering an image of the world linked to an itinerary (326–7):

> **We belong to the category of mammals that spend part of their existence inside an artificial shelter. In this respect we differ from the monkeys – among whom the most highly developed make only rough adjustments to the place where they will spend a night – but resemble the numerous rodents whose elaborately constructed burrows serve as the centre of their territory and often as their food store. [. . .] According to a deep-rooted scientific tradition, prehistoric humans lived in caves. If this were true, it would suggest interesting comparisons with the bear and the badger, omnivorous and plantigrade like ourselves, but it would be more correct to suppose that although humans sometimes took advantage of caves when these were habitable, they lived in the open in the statistically overwhelming majority of cases and, from the time when records become available, in built shelters. (Leroi-Gourhan, 1964, 318)**

we differ from the monkeys – among whom the most highly developed make only rough adjustments to the place where they will spend a night – but resemble the numerous rodents whose elaborately constructed burrows serve as the centre of their territory

There is the same assertion as in Deleuze and Guattari's work of continuity between animal and human behaviour, and an implication that mechanisms that were acquired at an earlier evolutionary stage can survive in us and can come into action when circumstances constitute them in an appropriate machine. Migratory birds are nomadic, but song birds are territorial. Monuments are refrains (Deleuze and Guattari, 1994, 184). Nature, including the man-made environment, is polyphonic. Each species has its own world, created through its percepts, concepts and affects; each recognizes its own patterns and has its own refrain; and their territories intersect. The spider, for example, seems to have something of the fly in it:

> **It has often been noted that the spider web implies that there are sequences of the fly's own code in the spider's code; it is as though the spider had a fly in its head, a fly 'motif', a fly 'refrain'. The implications may be reciprocal, as with the wasp and the orchid, or the snapdragon and the bumblebee. Jakob von Uexküll has elaborated an admirable theory of transcodings. He sees the components as melodies in counterpoint, each of which serves as a motif for another: Nature as music. (Deleuze and Guattari, 1980, 314)**

. . . the interdependent voices of singers and instruments – each with their individuality, but making a sublime whole – worked like a natural ecology.

Uexküll's contrapuntal account of nature was prompted by thinking about an encounter he had with an intense young man who was at the Concertgebouw in Amsterdam, listening to a Mahler symphony (I would guess that it was the third, given the forces it deploys and its subject-matter: 'Pan awakes', 'What the flowers of the field tell me', 'What the creatures of the forest tell me', etc.) completely absorbed in a copy of the musical score. Uexküll thought this an odd thing for the young man to be doing, when he could have been opening himself up to direct experience of the music, but the stranger explained that he could hear more of what was going on if he allowed his eyes to help him with information from the score. 'With the score one can follow the growth and ramification of individual voices, which launch themselves like the pillars of a cathedral's nave and carry the vault of the work' (Uexküll, 1934, 147). Uexküll saw that the interdependent voices of singers and instruments – each with their individuality, but making a sublime whole – worked like a natural ecology, and that his own task, as a biologist, was to write the score of nature. Deleuze and Guattari took up Uexküll's perception with enthusiasm:

> **On the death of the mollusc, the shell that serves as its house turns into the counterpoint of the hermit crab that turns it into its own habitat, thanks to its tail, which is not for swimming but is prehensile, enabling it to capture the empty shell. The tick is organically constructed in such a way that it finds its counterpart in any mammal whatever that passes below its branch, as oak leaves arranged in the form of tiles find their counterpoint in the raindrops that stream over them. This is not a teleological conception but a**

oak leaves arranged in the form of tiles find their counterpoint in the raindrops that stream over them. This is not a teleological conception but a melodic one in which we no longer know what is art and what nature

> **melodic one in which we no longer know what is art and what nature ('natural technique'). There is a counterpoint whenever a melody arises as a 'motif' within another melody, as in the marriage of bumblebee and snapdragon. These relationships of counterpoint join planes together, form compounds of sensations and blocs, and determine becomings. But it is not just these determinate *melodic compounds*, however generalized, that constitute nature; another aspect, an infinite *symphonic plane of composition*, is also required: from House to universe. From endosensation to exosensation. This is because the territory does not merely isolate and join but opens onto cosmic forces that arise from within or come from outside, and renders their effect on the inhabitant perceptible. (Deleuze and Guattari, 1994, 185).**

The importance of pattern recognition in music and nature are further explored in connection with emergent phenomena in Douglas Hofstadter's book *Gödel, Escher, Bach* (1979) 'a metaphorical fugue on minds and machines' and one of the foundational works on artificial intelligence. Johann Sebastian Bach (1685–1750) is – of course – the pre-eminent emblem for the idea of counterpoint, and Hofstadter's influence has been such that perhaps we are no longer surprised to find Bach drawn into the analysis of what it is to think. Uexküll's vision of nature, as endorsed by Deleuze and Guattari, infers an immanent and pervasive Bach-like sensibility in nature, or at least in nature's score; and we should not separate the human from the natural: Proust infers it in bourgeois society. The Baron de Charlus emboldens himself by humming a tune to himself as he sets off in pursuit of his next sexual encounter (with the tailor Jupien) though at this point in the text his purpose is not fully explicit. It becomes clearer in the next sentence, by turning the chase into a little fugue: 'At the same instant as M. de Charlus disappeared through the gate humming like a great bumble-bee, another, a real one this time, flew into the courtyard. For all I knew this might be the one so long awaited by the orchid, coming to bring it that rare pollen without which it must remain a virgin'. And then the metaphorical orchid reappears: 'I was distracted from following the gyrations of the insect, for, a few minutes later, engaging my attention afresh, Jupien [. . .] returned, followed by the Baron' (Proust, 1913–27, 4, 7). The refrain, *ritournelle*, returns repeatedly in a poignant 'little phrase' composed by the fictional character Vinteuil. As it is a fictional little tune, we have never actually heard it, but in the novel we come to recognize it in precisely a musical way when it returns and brings with it a trailing cloud of memories, which are, by the time it happens, our own as well as those of the books' characters. 'We require just a little order to protect us from chaos' (Deleuze and Guattari, 1994, 201). We are highly predisposed to recognize order and attach significance to it. We feel secure when there is some order of a kind that we recognize, and at the same time we discount everything else as being somehow beside the point.

We require just a little order to protect us from chaos.

We are surprisingly susceptible to 'conspiracy theories', which see an underlying order in unconnected events, and the paranoid sees order everywhere, organized so as to persecute him. There are times when chaos might seem to be on the point of overwhelming us, but if we have managed to regulate our lives in such a way that we have habits that help us to do the things we're trying to do, then these eruptions of chaos will be rare, and will be experienced as crises. If I miss my usual train, then I feel disappointed and inconvenienced; I might need to make a phone call, but I don't feel that chaos is upon us. But when, on my way to the station, I start to feel my body changing into a wolf's, then – whether I panic or not – I experience something much more like chaos; at least until I have worked out what is going on. Chaos in the Deleuze and Guattari world is a body without organs, the schizophrenic body, the plane of immanence, where things are forming and being taken apart as fast as they form. Emergent order is held at bay, and never emerges. A little order – a tune, a heartbeat – and the chaos recedes; a possibility emerges from a plateau of stability. Deleuze and Guattari's image of chaos is far from inert. It is continually making and unmaking:

> **Chaos is defined not so much by its disorder as by the infinite speed with which every form taking shape in it vanishes. It is a void that is not a nothingness but a *virtual*, containing all possible particles and drawing out all possible forms, which spring up only to disappear immediately, without consistency or reference, without consequence. Chaos is an infinite speed of birth and disappearance. (Deleuze and Guattari, 1991, 118)**

Orpheus and Ariadne

Deleuze and Guattari make a link between music and politics, and the idea does not originate with them. Orpheus was portrayed in ancient mythology as the person who founded Greece by playing music. He is depicted in the company of wild animals, who calmly listen to his music rather than kill one another; and this is said to represent him as the originator of the laws that drew city states into civilized harmony, rather than continuing in barbarous warfare (Ballantyne, 1997, 181). The song of Orpheus is the song of order being conjured out of chaos, the song of a legislator establishing territory, or to put it another way, it is birdsong.[3]

It has been said often enough that architecture is frozen music, and the remark is attributed to different people, though Schelling seems to have the claim to precedence.[4] Having established itself as cliché, it is well on the way to becoming self-evidently true, and the concept is embodied in thought more often and more anciently than the particular expression implies – for example in Renaissance proportion and in ancient Egyptian, Indian and Chinese creation myths (Pascha, 2004). The implied 'sonority' of Ariadne's thread is there because it too is an ordering principle. By making use of it, Theseus found his way out of the monstrous anti-architecture of the Minotaur's labyrinth, certainly a place to defeat music. With the thread, Theseus' confusions melted away and his path became clear. Nietzsche composed and wrote about music. He went through a phase where he admired Wagner, but went on from that position, later finding Wagner burdensome and, he said 'Germanic', whereas he (Nietzsche) wanted to endorse the exuberant, the light, the Dionysian and Greek. (I continue to use Nietzsche's terms below when presenting Nietzsche's text, but I have above been using 'commonsensical' to mean what Nietzsche meant by 'Germanic', and do not want to attach 'common sense' to any one nation in particular.) Nietzsche found affirmative music best exemplified in Bizet's *Carmen:* 'This music seems perfect to me,' he said (Nietzsche, 1888, 157). Deleuze presents Ariadne as moving away from her attachment to Theseus – ponderous and 'German' *avant la lettre* – to the life-affirming joyous Dionysus. At which point her world turns inside out, and ours along with it. No longer does the labyrinth look confusing. It is transmuted into an image of all that is how it should be:

> **The labyrinth is no longer architectural; it has become sonorous and musical. It was Schopenhauer who defined architecture in terms of two forces, that of bearing and that of being borne, support and load, even if the two tend to merge together. But music appears to be the very opposite of this, when Nietzsche separates himself more and more from the old forger, Wagner the magician: music is Lightness [*la Legère*]. Pure weightlessness.[5] Does not the entire triangular story of Ariadne bear witness to an anti-Wagnerian lightness, closer to Offenbach and Strauss than to Wagner? To make the roofs dance, to balance the beams – that is what is essential to Dionysus the musician.[6] Doubtless there is also an Appolonian, even Theseusian side to music, but it is a music that is distributed according to territories, *milieus*, activities ethoses: a work song, a marching song, a dance song, a song for repose, a drinking song, a lullaby . . . almost little**

> 'hurdy-gurdy songs', each with its particular weight.[7] In order for music to free itself, it will have to pass over to the other side – there where the territories tremble, where the structures collapse, where the ethoses get mixed up, where a powerful song of the earth is unleashed, the great *ritornello* that transmutes all the airs it carries away and makes return.[8] *Dionysus knows no other architecture than that of routes and trajectories.* Was this not already the distinctive feature of the *lied*: to set out from the territory at the call or wind of the earth? Each of the higher men leaves his domain and makes his way towards Zarathustra's cave. But only the dithyramb spreads itself out over the earth and embraces it in its entirety. Dionysus has no territory because he is everywhere on the earth.[9] The sonorous labyrinth is the song of the earth, the *Ritornello*, the eternal return in person. (Deleuze, 1993, 104)

The text is highly allusive. Deleuze's footnotes take us to the allusions in Nietzsche's texts. There is a contrast here between the German *lied* and the Greek *dithyramb*. The '*ritornello*' is an alternate translation of Deleuze and Guattari's *ritournelle*, which has elsewhere been turned into the 'refrain', but here crucially its 'returning' character is linked with, even conflated with, Nietzsche's 'eternal return' – an important concept in Nietsche's writings, which in Deleuze's interpretation is the ever-recurring creative moment, when one thinks, or acts with genuine spontaneous impulses in response to the circumstances one finds oneself to be in. It is the eternal return of those moments when one feels oneself to be most alive, rather than doing one's duty or giving back an answer that one has been taught earlier. It is what I think Winckelmann was trying to describe when he sought to explain, in a passage that was certainly known to Nietzsche, the peculiar genius of the ancient Greeks. 'Behold the swift Indian outstripping in pursuit the hart: how briskly his juices circulate! How flexible, how elastic his nerves and muscles! How easy his whole frame! Thus Homer draws his heroes' (Winckelmann, 1755, 6). Winckelmann portrayed the Greeks as being formed by their circumstances, but then acting spontaneously and responding directly to life's stimuli. This spontaneous 'Greek' engagement with the world contrasts – in Nietzsche's idiom – with the 'German', which represents the modern everyday world with its commercial practical outlook, philistine and dull – the world of getting things done on time, machine-produced hurdy-gurdy music helping to move things along. It is the world of common sense; and it could hardly be further from the world view evoked by Gustav Mahler, whose Austrian nationality should not make us think that he falls into Nietzsche's category of 'the Germanic'. He wrote *Das Lied von der Erde* (*The Song of the Earth,* 1908). 'Territory [*La territoire*] is German, but the Earth [*la Terre*] is Greek,' said Deleuze and Guattari:

> And this disjunction is precisely what determines the status of the romantic artist, in that she or he no longer confronts the gaping chaos but the pull of the Ground [*attirance du Fond* – the pull of the deep]. The little tune, the bird's refrain, has changed: it is no longer the beginning of a world but draws a territorial assemblage upon the earth [*elle trace sur la terre l'agencement territorial*]. It is then no longer made of two consonant parts that seek and answer one another; it addresses itself to a deeper singing that founds it [*un chant plus profond qui la fonde*], but also strikes against it and sweeps it away, making it ring dissonant. The refrain is indissolubly constituted by the territorial song and the song of the earth that rises to cover it. Thus at the end of *The Song of the Earth* two motifs coexist, one melodic, evoking the assemblages of the bird, the other rhythmic, evoking the deep breathing of the earth, eternally. Mahler says that the singing of the birds, the colour of the flowers, and the fragrance of the forest are not enough to make Nature, that the god Dionysus and the great Pan are needed. The Ur-refrain of the earth harnesses all refrains whether territorial or not, and all *milieu* refrains. (Deleuze and Guattari, 1980, 339, translation modified; v.o., 418).

Consolidation

One of the themes that emerges in Deleuze and Guattari's work is the explanatory power of environments. There will be more to say about environments in the chapters that follow, but they point out, for example that

> The philosopher Eugène Dupréel [1879–1967] proposed a theory of *consolidation*; he demonstrated that life went not from a centre to an exteriority but from an exterior to an interior, or rather from a discrete or fuzzy aggregate to its consolidation. This implies three things. First, that there is no beginning from which a linear sequence would derive, but rather densifications, intensifications, reinforcements, injections, showerings, like so many intercalary events ('there is growth only by intercalation'). Second,

and this is not a contradiction, there must be an arrangement of intervals, a distribution of inequalities, such that it is sometimes necessary to make a hole in order to consolidate. Third, there is a superposition of disparate rhythms, an articulation from within of an interrhythmicity, with no imposition of meter or cadence.[10] Consolidation is not content to come after; it is creative. The fact is that the beginning always begins in-between, intermezzo. Consistency is the same as consolidation, it is the act that produces consolidated aggregates, of succession as well as of coexistence, by means of three factors just mentioned; intercalated elements, intervals, and articulations of superposition. Architecture, as the art of the abode and the territory, attests to this: there are consolidations that are made afterward, and there are consolidations of the keystone type that are constituent parts of the ensemble. More recently, matters like reinforced concrete have made it possible for the architectural ensemble to free itself from arborescent models employing tree-pillars, branch-beams, foliage-vaults. Not only is concrete a heterogeneous matter whose degree of consistency varies according to the elements in the mix, but iron is intercalated following a rhythm; moreover, its *self-supporting surfaces* form a complex rhythmic personage whose 'stems' have different sections and variable intervals depending on the intensity and direction of the force to be tapped (armature instead of structure). In this sense, the literary or musical work has an architecture: 'Saturate every atom,' as Virginia Woolf said;[11] or in the words of Henry James, it is necessary to 'begin far away, as far away as possible,' and to proceed by 'blocks of wrought matter'. It is no longer a question of imposing a form upon a matter but of elaborating an increasingly rich and consistent material, the better to tap increasingly intense forces. (Deleuze and Guattari, 1980, 328–9)

'Intercalation' has a variety of meanings, including a geological one which describes a stratum of a substance being found between strata of something quite different and distinct. Reinforcing-steel is intercalated in concrete in Deleuze and Guattari's example above, and similarly architecture is intercalated into life, structuring and framing it. Virginia Woolf's 'atoms' were atoms of experience, which she was trying to include in her literary compositions. She was attempting, she said:

To give the moment whole; whatever it includes. Waste, deadness, come from the inclusion of things that don't belong to the moment; this appalling narrative business of the realist: getting on from lunch to dinner: it is false, unreal, merely conventional. Why admit any thing to literature that is not poetry – by which I mean saturated? [. . .] The poets succeed by simplifying: practically everything is left out. I want to put practically everything in, yet to saturate.' (Woolf, 1980, vol 3, 209–10, 28 November 1928)

One can see a parallel between the attitude expressed here and the way that Deleuze and Guattari decline to spell out every step of an argument. The allusions to Henry James refer to a passage in the preface that he added to *The Wings of the Dove* in 1909 in which he draws on an idea of building. He proposes a stereotomy of the novel, or a stereotomy of the *character* in the novel – something cut from the solid, and robustly built. James here is discussing the way he sought to build his two principal female protagonists, Milly Theale and Kate Croy, so that when they carried out the actions that the story demands of them, the reader would find the actions plausible as things the characters might do. Music and masonry fuse together in James's language:

The poets succeed by simplifying: practically everything is left out. I want to put practically everything in, yet to saturate.'

The great point was, at all events, that if in a predicament she was to be, accordingly, it would be of the essence to create the predicament promptly and build it up solidly, so that it should have for us as much as possible its ominous air of awaiting her. That reflexion I found, betimes, not less inspiring than urgent; one begins so, in such a business, by looking about for one's compositional key, unable as one can only be to move till one has found it. (James, 1909, x)

Preparatively and, as it were, yearningly – given the whole ground – one began, in the event, with the outer ring, approaching the centre thus by narrowing circumvallations. [. . .] I felt this perfectly, I remember, from the moment I had comfortably laid the ground provided in my first Book, ground from which Milly is superficially so absent. I scarce remember perhaps a case [. . .] in which the curiosity of 'beginning far back,' as far back as possible, and even of going, to the same tune, far 'behind,' that is behind the face of the subject, was to assert itself with less scruple. (James, 1909, xi)

[Publication demands compromises, to which the craftsman-writer can respond with ingenuity.] The best and finest ingenuities, nevertheless, with all respect to that truth, are apt to be, not one's compromises, but one's fullest conformities, and I well remember, in the case before us, the pleasure of feeling my divisions, my proportions and general rhythm, rest all on permanent rather than in any degree on momentary proprieties. It was enough for my alternations, thus, that they were good in themselves; it was in fact so much for them that I really think any further account of the constitution of the book reduces itself to a just notation of the law they followed.

the pleasure of feeling my divisions, my proportions and general rhythm, rest all on permanent rather than in any degree on momentary proprieties.

There was the 'fun,' to begin with, of establishing one's successive centres – of fixing them so exactly that the portions of the subject commanded by them as by happy points of view, and accordingly treated from them, would constitute, so to speak, sufficiently solid BLOCKS of wrought material, squared to the sharp edge, as to have weight and mass and carrying power; to make for construction, that is, to conduce to effect and to provide for beauty. Such a block, obviously, is the whole preliminary presentation of Kate Croy, which, from the first, I recall, absolutely declined to enact itself save in terms of amplitude. Terms of amplitude, terms of atmosphere, those terms, and those terms only, in which images assert their fullness and roundness, their power to revolve, so that they have sides and backs, parts in the shade as true as parts in the sun [. . .] . I have just said that the process of the general attempt is described from the moment the 'blocks' are numbered, and that would be a true enough picture of my plan. Yet one's plan, alas, is one thing and one's result another; so that I am perhaps nearer the point in saying that this last strikes me at present as most characterized by the happy features that were, under my first and most blest illusion, to have contributed to it. [. . .] The building-up of Kate Croy's consciousness to the capacity for the load little by little to be laid on it was, by way of sufficiently solid BLOCKS of wrought material, squared to the sharp edge, as to have weight and mass and carrying power; to make for construction, that is, to conduce to effect and to provide for beauty.

example, to have been a matter of as many hundred close-packed bricks as there are actually poor dozens. (James, 1909, xii-xiv)

The blocs of experience and sensation that shape a person's development, are turned into an image of building-blocks that construct the novelist's characters, and they develop with reference to rhythms and amplitudes. The novelist simplifies of course, but one of the reasons we respond to James's characters is that we feel as though we know them. We have witnessed various formative experiences along with them, so to some extent we vicariously share their intuitions. It is a lesson that the film industry has brilliantly learnt, so that we can emote at the same time as the characters on the screen. Part of the skill in screen acting is being able to look blank enough for the audience to project its emotions on to the actor, and part of the editor's skill is knowing how long the audience needs. The mainstream film *Thelma and Louise* (1991) is a story about deterritorialization, where the characters are taken out of their common-sense routines, have various life-enhancing and traumatic experiences, which we witness with them, and they end by embracing death by driving off a cliff into the Grand Canyon. The canyon is the culmination of a series of breathtaking landscapes, including Monument Valley, where the majesty of the earth in desert conditions is evoked by the combination of images and music. After the desert has worked on the women, it seems preferable to them to choose death – absolute deterritorialization – to going back home to resume the old routines. There is a moment when Thelma's old life territorializes around the sound of her husband's voice on the telephone, and we feel it trying to take hold of her again; but she resists it and breaks off the conversation. The story could have been told as a descent into madness and despair, but we are taken through the characters' experiences in a way that allows us to feel, with them, that suicide is the optimistic option. So the film ends with a freeze-frame on their car, after it has driven off the cliff, but before it has started to fall. The montage of their happiest moments together, and the up-beat music, leave one with the impression that deterritorialization is the answer. In this case it is the sound of the husband's voice that is the territorializing tune, that structures Thelma's mental space into the frame of the domestic routines that used to shape her life. The clutter of terrible furniture in the house, and the husband's assumption that he has the right to issue orders to her, contrasts with her freedom and the profundity of emotions she has experienced on the road as her horizons have broadened. Establishing territory is architecture's great and normal role. The monument is a song. A building usually establishes a practical domain, and often marks out the extent of a proprietor's property, but aside from establishing ownership, the territory it marks out is a zone where a certain ethos applies: a work place, a drill ground, a dance hall, a quiet hotel lounge, a convivial bar, a cocooned bedroom . . . almost little 'hurdy-gurdy places'. The architecture helps us to do the things that need to be done, and reinscribes the established order. The clutter stops one seeing beyond it. This is the architecture of Thelma before her escape. Or the architecture of Theseus, the heroic princely embodiment of the ordinary jock, who shows physical courage and ingenuity, who has spent time in the gym and has put on the muscle that will defeat a superior being. He never understands the labyrinth or the Minotaur, but he outwits them by a ruse – the thread supplied by Ariadne in her 'cheerleader' days. Establishing territories in this way is a necessary developmental phase, and it is the role that buildings normally have. 'Art begins not with flesh, but with the house. That is why architecture is the first of the arts' (Deleuze and Guattari, 1994, 186). Having established a house, one can take steps outside it – towards an architecture where the territories tremble, where the ethoses get mixed up, but it seems to be more like work on oneself than on buildings – each of the higher men *leaves* his domain – the structures collapse. Dionysus knows no other architecture than that of routes and trajectories. He has no territory because he is everywhere on the earth.

Dionysus knows no other architecture than that of routes and trajectories.

Desertification: Thelma phones home. Geena Davis as Thelma Yvonne Dickinson in *Thelma and Louise*, directed by Ridley Scott, 1991.

The domestic regime reterritorializes. Christopher McDonald as Darryl Dickinson on the telephone in *Thelma and Louise*, directed by Ridley Scott, 1991.

Against the sky: off road in *Thelma and Louise*, directed by Ridley Scott, 1991.

House, earth, territory

Buildings act as part of machines. The building-object is part of a machine that is activated and becomes productive when it is in use; and a single building-object, even something as simple as a little hut, might be taken up and used in different ways at different times or by different groups of people. Then it would become part of different machines and would produce something different in each case. What buildings produce most often is a territory – a space where a particular order prevails or seems implicit. A building is a little song. If it is a territory that is produced for the sake of achieving a particular end, then the song is a rather mechanical song – a work song, a marching song, a hurdy-gurdy song – that helps us to get things done without engaging us in various ways that we could be engaged, but which would actually be unhelpful in everyday circumstances. Commercial places know the value of hurdy-gurdy songs: the loud pop music that signals that the clothes for sale here will appeal to the under-25s, the educated-sounding classical music that plays in bookshops, the weirdly soporific music that plays in hotel lifts. These little songs establish little territories, and architecture can help them on their way; but this is an architecture of small horizons. Architecture can open to other possibilities, which are introduced here: there is the great 'song of the earth', which resonates through everything, and there is the architecture of trajectories, where buildings seem to dissolve away with the dissolution of the territories that become unnecessary to the all-pervasive Dionysus. Most of us, most of the time, want to feel secure in the territory that we know and welcome as our own, putting us in the position of the hefted sheep or the twittering birds. But at the important moments in our lives, the moments when we are most fully alive, we must pay attention to the deep resonance that the earth asserts everywhere, through all territories, or the disorienting freedoms that keep us moving through new unstable spaces that open up new possibilities, however incomprehensible and unproductive they might seem when we are operating in the world of common sense. At those moments, the voice of everyday reason can sound so oppressive and limiting that the only thing to do is to hang up the phone.

CHAPTER 4

Façade and Landscape

A walk in the mountains

Opening up to the outside, where things are different from the territory we know and inhabit, has its dangers. If we make it a constant habit then we might find that we have lost any sense of who we are, and have become schizophrenic. In *Capitalism and Schizophrenia* Deleuze and Guattari embrace this tendency. For example, Jakob Lenz (1751–92) who wanders through the opening pages of *Anti-Oedipus*, finds himself involved with his surroundings to an extraordinary degree. He was born in Livonia (now Latvia) as part of its German elite, and he went to study in Germany, where he fell in with Goethe and the group of Romantic poets belonging to the Sturm und Drang group. Nowadays he is remembered principally as a literary creation by Georg Büchner (1813–1837) who was born two generations later, but whose brilliant reimagining of Lenz's life between 20 January and 8 February 1778 has become part of the modernist canon of German literature. Goethe moved in aristocratic circles, but as Lenz's mental afflictions grew more pronounced, he felt unable to cope, and he exiled himself to more rural settings before eventually settling in Moscow, where he lived during the years after his obituary had been published in Germany (Sieburth, 2004). His short stay with a pastor, Johann Friedrich Oberlin, in Waldersbach in the Vosges mountains, was documented by Oberlin himself, and he certainly found his visitor startling, but his descriptions of Lenz's actions, while respectful and anxious, simply see them as bizarre. On the first night, for example, Lenz disturbed the neighbours in the middle of the night by climbing into the fountain and splashing about in it like a duck (Oberlin, 1778, 85). Büchner gives a very different account of the events, trying to describe them, as it were, from inside Lenz's mind. So as Lenz in the narrative is approaching Waldbach, in a single extraordinary sentence, Büchner evokes an exhilarating confusion of images of a place that is insecurely grasped, but which is experienced with great intensity:

> Only sometimes when the storms tossed the clouds into the valleys and they floated upwards through the woods and voices awakened on the rocks, like far-echoing thunder at first and then approaching in strong gusts, sounding as if they wanted to chant the praises of the earth in their wild rejoicing, and the clouds galloped by like wild whinnying horses and the sunshine shot through them and emerged and drew its glinting sword on the snowfields so that a bright blinding light knifed over the peaks into the valleys; or sometimes when the storms drove the clouds downwards and tore a light blue lake into them and the sound of the wind died away and then like the murmur of a lullaby or pealing bells rose up again from the depths of ravines and tips of fir trees and a faint reddishness climbed into the deep blue and small clouds drifted by on silver wings and all the mountain peaks, sharp and firm, glinted and gleamed far across the countryside, he would feel something tearing at his chest, he would stand there, gasping, body bent forward, eyes and mouth open wide, he was convinced he should draw the storm into himself, contain everything within himself, he stretched out and lay over the earth, he burrowed into the universe, it was a pleasure that gave him pain; or he would remain still and lay his head upon the moss and half-close his eyes and then everything receded from him, the earth withdrew beneath him, it became as tiny as a wandering star and dipped into a rushing stream whose clear waters flowed beneath him. But these were only moments, and then he got up, calm, steady, quiet, as if a shadow play had passed before him, he had no memory of anything. (Büchner, 1839, 3–8)

This Lenz is thoroughly deterritorialized, at least for the duration of these 'shadow plays', and he certainly hears and knows how to respond to the song of the earth, even though it does not make much sense to those around him. Deleuze and Guattari draw attention to a spatial contrast – a move from confinement to expansiveness – from a passage where Lenz speaks with Oberlin in an enclosed room, to the contrast when Lenz then goes for a walk outside. When Lenz is with the pastor the conversation is controlled in such a way that he is allowed to situate himself only in relation to his father and mother – which is to say that he is kept in the Oedipalized relations of the family. The confinement of the room correlates with the oppression in his mind. The assemblage of the territorialized room and the pastor who makes that territorialization inescapable makes a machine that is uncomfortably powerful for Lenz's mind which is inclined to deterritorialize:

> While taking a stroll outdoors, on the other hand, he is in the mountains, amid falling snowflakes, with other gods or without any gods at all, without a family, without a father or a mother, with nature. 'What does my father want? Can he offer me more than that? Impossible. Leave me in peace.' Everything is a machine. Celestial machines, the stars or rainbows in the sky, alpine machines – all of them connected to those of his body. The continual whirr of machines. 'He thought that it must be a feeling of endless bliss to be in contact with the profound life of every form, to have a soul for rocks, metals, water, and plants, to take into himself, as in a dream, every element of nature, like flowers that breathe with the waxing and waning of the moon.' To be a chlorophyll- or a photosynthesis-machine, or at least slip his body into such machines as one part among the others. Lenz has projected himself back to a time before the man-nature dichotomy, before all the co-ordinates based on this fundamental dichotomy have been laid down. He does not live nature as nature, but as a process of production. There is no such thing as either man or nature now, only a process that produces the one within the other and couples the machines together. Producing-machines, desiring-machines everywhere, schizophrenic machines, all of species life: the self and the non-self, outside and inside, no longer have any meaning whatsoever. (Deleuze and Guattari, 1972, 2)

There is no such thing as either man or nature now, only a process that produces the one within the other and couples the machines together.

Lenz's sense of himself is completely absorbed into the landscape, or the elements of it. There is no sense of discontinuity, no separation, no borders between himself and his surroundings. Sensing the profound life of every form, he is in a position to give voice to the song of the earth, or a fragment of it, through his body or – if he can translate it into disciplined words – in his poems. Deleuze and Guattari correlate this vision of the world through the eyes of a schizophrenic with various things, including a general tendency to project an image of oneself into landscapes (and anything else, come to that) and with the shift in consciousness between sedentary and nomadic cultures. One of the ways in which we make sense of the world is to see ourselves reflected in it. We suppose that the world is in some ways like ourselves, and much of the time we are wrong, but it gets us by. The other side of the comparison necessarily is that we must be like the world around us. The English poet Lord Byron (1788–1824), who was no less a Romantic than Lenz, describes in his epic verse *Childe Harold*, his sense of his identity as something volatile that bonds with his changing surroundings. He changes and becomes a different sort of person in the different surroundings that he puts himself in, and sometimes this is liberating and thrilling, whereas at other times it is constraining and unwelcome:

> **I live not in myself, but I become**
> **Portion of that around me; and to me**
> **High mountains are a feeling, but the hum**
> **Of human cities torture. I can see**
> **Nothing to loathe in nature, save to be**
> **A link reluctant in a fleshly chain,**
> **Classed among creatures, when the soul can flee,**
> **And with the sky, the peak, the heaving plain**
> **Of ocean, or the stars, mingle and not in vain.**
> **(Byron, 1812–18, Canto 3, stanza 72)**

This is the same vision as Büchner's Lenz had, but here held in check because Byron felt himself unable (while still an embodied creature) to do what he longed to do and mingle with the sky, the mountains, the ocean and the stars, whereas Lenz was already doing just that.

White wall, black hole

With Deleuze and Guattari the most important correspondence in dealing with the landscape is the correspondence with the face, which reappears continually in fantastically varied guises. Their 'face' is made up of two components: the white wall and the black hole. The white wall is a reflective screen, which reflects back any information that is projected on to it. The black hole is the opposite principle – it reflects nothing at all, but absorbs everything into it. In the image of the face the pupil of the eye is this black hole – the traditional 'window of the soul' – but it need not necessarily be visible as such. The important thing is that there is something unknowable behind the white wall, a 'subject' with thoughts and feelings, which, if they are to be inferred at all, can be inferred only from the signs that are inscribed in or projected on the white wall. The 'face' is this dual operation of exclusion and absorption, reflection and reception. For example the white whale, Moby-Dick, is a white wall for the madness of Captain Ahab, who projects on to the whale an obsession that is entirely his own, but which the whale reflects back:

> **Ahab had cherished a wild vindictiveness against the whale, all the more fell for that in his frantic morbidness he at last came to identify with him, not only all his bodily woes, but all his intellectual and spiritual exasperations. The White Whale swam before him as the monomaniac incarnation of all those malicious agencies which some deep men feel eating in them, till they are left living on with half a heart and half a lung. [. . .] All that most maddens and torments; all that stirs up the lees of things; all truth with malice in it; all that cracks the sinews and cakes the brain; all the subtle demonisms of life and thought; all evil, to crazy Ahab, were visibly personified, and made practically assailable in Moby-Dick. He piled upon the whale's white hump the sum of all the general rage and hate felt by his whole race from Adam down; and then, as if his chest had been a mortar, he burst his hot heart's shell upon it. (Melville, 1851, chapter 41)**

The fact of the white whale's indifference to Ahab is not, for Ahab, a plausible conjecture. Ahab's previous encounter with the whale was traumatic. He lost his leg to it and endured great pain, during which he lost all sense of himself and became pure intensity – a body without organs. When he recovered his senses his identity reterritorialized around the whale, and now in the time of the story he so hates the whale that tore his leg off that he projects his own hatred on to the whale and stalks the whale, and comes to feel that the whale is stalking him, driven by the obsessive malign thoughts that the whale's white hump indifferently reflected back to him. It develops in his mind into the embodiment of every evil. 'All visible objects,' said Ahab,

> are but as pasteboard masks. But in each event – in the living act, the undoubted deed – there, some unknown but still reasoning thing puts forth the mouldings of its features from behind the unreasoning mask. If man will strike, strike though the mask! How can the prisoner reach outside except by thrusting through the wall? To me, the white whale is that wall, shoved near to me. Sometimes I think there's naught beyond. But 'tis enough. He tasks me; he heaps me; I see in him outrageous strength, with an inscrutable malice sinewing it. That inscrutable thing is chiefly what I hate; and be the white whale agent, or be the white whale principal, I will wreak that hate upon him. [. . .] I'd strike the sun if it insulted me. For could the sun do that, then could I do the other; since there is ever a sort of fair play herein. (Melville, 1851, chapter 36; quoted in part by Deleuze and Guatttari, 1980, 245).

'Captain Ahab,' say Deleuze and Guattari, 'is engaged in an irresistible becoming-whale with Moby-Dick; but the animal Moby-Dick must simultaneously become an unbearable pure whiteness, a shimmering pure white wall' (Deleuze and Guattari, 1980, 304). The shimmering pure white wall is a cinema screen, but not only that. It is also the 'screen of dreams', on to which our dream images are arranged, which is apparently to be identified with an infantile memory of the breast that, close-up, fills the field of vision (Deleuze and Guattari, 1980, 169). The condition might be approached in later life when close-ups of faces become landscapes on the cinema screen. 'Our work,' says Ingmar Bergman, 'begins with the human face. . .. The possibility of drawing near to the human face is the primary originality and the distinctive quality of the cinema' (Bergman in *Cahiers du cinéma*, October 1959, quoted by Deleuze, 1983, 99):

> Face and landscape manuals formed a pedagogy, a strict discipline, and were an inspiration to the arts as much as the arts were an inspiration to them. Architecture positions its ensembles – houses, towns or cities, monuments or

factories – to function like faces in the landscape they transform. Painting takes up the same movement but also reverses it, positioning a landscape as a face, treating one like the other: 'treatise on the face and landscape'. The close-up in film treats the face primarily as a landscape; that is the definition of film, black hole and white wall, screen and camera. But the same goes for the earlier arts, architecture, painting, even the novel: close-ups animate and invent all of their correlations. So, is your mother a landscape or a face? A face or a factory? (Godard.) All faces envelop an unknown, unexplored landscape; all landscapes are populated by a loved or dreamed-of face, develop a face to come or already past. What face has not called upon the landscapes it amalgamated, sea and hill; what landscape has not evoked the face that would have completed it, providing an unexpected complement for its lines and traits? (Deleuze and Guattari, 1980, 173)

It is not crucial for the argument, but is nevertheless significant in the way that it is made, that in French the words for face (*visage*) and landscape (*paysage*) are euphonious, and sound as if they are related, along with Deleuze and Guattari's new words, developed from them – *visagéité* and *paysagéité* – which translate into English as 'faciality' and 'landscapicity', much more uncomfortable words, which are nevertheless useful in this context. Deleuze and Guattari quote Chrétien de Troyes' *Legend of the Grail*, which he wrote in the late twelfth century:

The novel – Perceval saw a flight of geese that had been dazzled by the snow. (. . .) A falcon had found one that had become separated from the flock. It attacked and struck her so hard that she fell to the earth; (. . .) When Perceval saw the disturbed snow where the goose had lain, with the blood still visible, he leaned upon his lance to gaze at this sight for the blood mingled with the snow resembled the blush of his lady's face. He became lost in contemplation: the red tone of his lady's cheeks in her white face were like the three drops of blood against the whiteness of the snow. As he gazed upon this sight, it pleased him so much that he felt as if he were seeing the fresh colour of his fair lady's face. (. . .) We have seen a knight asleep upon his charger. (Deleuze and Guattari, 1980, 173)[1]

And they continue, now in their own voice:

> **Everything is there: the redundancy specific to the face and landscape, the snowy white wall of the landscape-face [*paysage-visage*], the black hole of the falcon and the three drops distributed on the wall; and, simultaneously, the silvery line of the landscape-face spinning toward the black hole of the knight deep in catatonia. Cannot the knight, at certain times and under certain conditions, push the movement further still, crossing the black hole, breaking through the white wall, dismantling the face – even if the attempt may backfire. (Deleuze and Guattari, 1980, 173)**

Not only is Chrétien's image striking, it is also markedly cinematic. It powerfully calls to mind an image from the closing scenes of *Dangerous Liaisons* (1988) where, after a duel in which the Viscomte de Valmont has, rather to his surprise, been mortally wounded, there is a cut to an aerial view, a plan, and we see the scene of carnage from above – the black clothes and deep red blood along with the scars and trails left in the pristine surrounding snow, the victorious but naïve Chevalier Danceny catatonic while the manservant gently moves Valmont's clothing (see below). The point here is not so much that the scene looks like a face, but that it operates in the way that

White Snow, dark clothes, a trail of deep red blood, in *Dangerous Liaisons*, directed by Stephen Frears, 1988.

a face operates, by establishing a screen and then suggesting by way of marks and symbols on the screen that there is something behind it finding expression. The whiteness of the snowy ground carries traces of blood and dark marks made by disrupting the surface, and they are signs that express what has been going on. It also resonates with the subsequent image of the face of the Marquise de Merteuil, who, despite being absent from the scene, is the direct and immediate cause of this carnage. At the moment of the duel, and of his death, Valmont is the only person who understands this, but as he dies he hands over to his assassin the means of destroying the marquise. The film's closing image shows her wiping away her make-up and she divests herself of her *persona* (see below). Her carefully constructed identity has already been annihilated; her mask follows. Her make-up whitens her already pale face, and gives her deep blood-red lips, while her eyes, already darkened with mascara, are in shadow. Her face is the scene of carnage; the scene of carnage is her face. The images infuse one another just as in Chrétien's image, and the face as the red and the white intensifiers of signification are wiped away is expressionless and dead. It becomes absolutely a white wall on to which we, the audience, can project our feelings about what the character might be thinking and feeling.

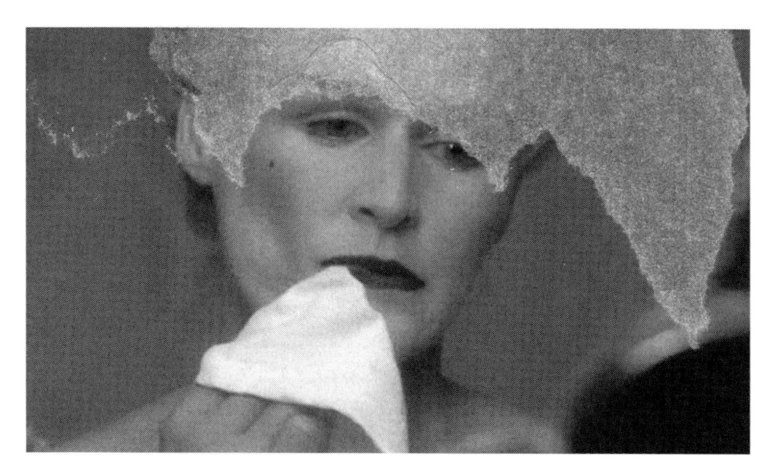

Whitened skin, darkened eyes, blood-red lips, Glenn Close as the Marquise de Merteuil , in *Dangerous Liaisons*, directed by Stephen Frears, 1988.

There is no longer anything coming from within, only our own feelings reflected back to us.

Deleuze and Guattari mention another example: Malcolm Lowry's *Ultramarine*, in which, in a scene dominated by the 'machinery' of the boat:

> **a pigeon drowns in waters infested by sharks, 'as if a red leaf should fall on a white torrent' (Lowry, 1933, 170) and this inevitably evokes the image of a bloody face. Lowry's scene is embodied in such different elements and is so particularly organized that there can be no question of influence by Chrétien de Troyes's scene, only confluence with it. This makes it an even better confirmation of the existence of a veritable black hole or red mark-white wall abstract machine (snow or water). (Deleuze and Guattari, 1980, 533, note 8)**

So the concept of the face, once it is established, is highly mobile and adaptable. We see it everywhere. However, it is not universal. At some point in the past it was a concept that we acquired, and it is a concept that the schizophrenic can lose.

> **Dismantling the face is no mean affair. Madness is a definite danger: Is it by chance that schizos lose their sense of the face, their own and others', their sense of the landscape, and the sense of language and its dominant signification all at the same time? The organization of the face is a strong one. (Deleuze and Guattari, 1980, 188)**

Nevertheless the face is not found in all cultures. The mask-face that gives the impression of detaching itself from the head behind it is a concept that Deleuze and Guattari identify specifically with Jesus Christ, and cultures that have had no contact with 'White Man' have found other ways to deal with the world. Their masks can emphasize the head as part of the body, rather than the face as a free-floating phantasm. These 'probe-heads' conceptualize the world and deal with it without ever drawing on this landscape-face that to us seems so pervasive and unavoidable (Deleuze and Guattari, 1980, 176). A certain type of power-relation becomes possible by means of the face – 'the maternal power operating through the face during nursing; the passional power operating through the face of the loved one, even in caresses; the political power operating through the face of the leader (streamers, icons, and photographs), even in mass actions; the power of film operating through the face of the star and the close-up; the power of television' (Deleuze and Guattari, 1980, 175). There is the 'four-eye machine' made of elementary faces linked two by two. 'The face of a teacher and a student, father and son, worker and boss, cop and citizen, accused and judge ('the judge wore a stern expression, his eyes were horizonless . . .'): concrete individualized faces are produced and transformed on the basis of these units, these combinations of units – like the face of a rich child in which a military calling is already discernible, that West Point chin. You don't so much have a face as slide into one' (Deleuze and Guattari, 1980, 177). Deleuze and Guattari see the example of Christ as a crucial moment in the development of the face, as the wounded body is facialized: 'Not only did Christ preside over the facialization of the entire body (his own) and the landscapification of all milieus (his own), but he composed all of the elementary faces and had every divergence at his disposal' (Deleuze and Guattari, 1980, 178). Back in 1946 Deleuze had published an essay, apparently his first published article, 'From Christ to the Bourgeoisie', which saw some such practice as a precondition for capitalism, and the argument is taken as read in *A Thousand Plateaus*. The body is overcoded, reterritorialized; a new 'subject' is constructed:

> there is no signifiance without a despotic assemblage, no subjectification without an authoritarian assemblage, and no mixture between the two without assemblages of power that act through signifiers and act upon souls and subjects. It is these assemblages, these despotic or authoritarian formations, that give the new semiotic system the means of its imperialism, in other words, the means both to crush the other semiotics and protect itself against any threat from outside. A concerted effort is made to do away with the body and corporeal coordinates through which the multidimensional or polyvocal semiotics operated. Bodies are disciplined, corporeality dismantled, becomings-animal hounded out, deterritorialization pushed to a new threshold – a jump is made from the organic strata to the strata of signifiance and subjectification. A single substance of expression is produced. The white wall/black hole system is constructed, or rather the abstract machine is triggered that must allow and ensure the almightiness of the signifier as well as the autonomy of the subject. [. . .], with the difference between our uniforms and clothes and primitive paintings and garb is that the former effect a facialization of the body, with buttons for black holes against the

> **white wall of the material. Even the mask assumes a new function here, the exact opposite of its old one. For there is no unitary function of the mask, except a negative one (in no case does the mask serve to dissimulate, to hide, even while showing or revealing). Either the mask assures the head's belonging to the body, its becoming-animal, as was the case in primitive societies. Or, as is the case now, the mask assures the erection, the construction of the face. The inhumanity of the face. (Deleuze and Guattari, 1980, 180–81)**

The point of the uniforms is not the buttons, but the fact that the clothes have taken on a role in the realm of signifiance. In very primitive asocial conditions we might wear clothes for the sake of keeping warm or to keep the sun off, but as soon as others are involved other functions for clothes come into play, such as ideas of concealment and exposure, the play of decency and decorum. But normally in our everyday activities these functions for clothing are not the issue that concerns us. We are more concerned with the signals the clothes send out, and they can be finely nuanced, and are not restricted to the apparel we put on in the morning. The streaked hair and the sun-bed tan are clothing, and tell us something about the person wearing them. 'What is the beauty of a building to us today?' asked Nietzsche, who answered himself by saying that it was 'the same thing as the beautiful face of a mindless woman: something mask-like' (Nietzsche, 1878, 218). We wear our buildings like we wear our uniforms. Artists wear black. Accountants wear particular sorts of stripes. Conservationists wear tweed. Architects wear black when they want to appear creative, stripes when they need to be trusted with large sums of money, and tweed when they are dealing with historic buildings. Buildings wear the façades they need to convey to the external world what they are and who they accommodate. 'Leroi-Gourhan established a distinction and correlation between two poles, 'hand-tool' and 'face-language' [. . .] it was a question of distinguishing a form of content and a form of expression' (Deleuze and Guatttari, 1980, 302). It makes a distinction between doing things and communicating things, which ultimately cannot be sustained as a clear division, but it is a separation that in some circumstances has its uses. There is a distinction in Uexküll between an animal's *Merkwelt* – its perceptual world, as apprehended through its organs of sensation (*le monde noté*) and its *Wirkwelt* its world of actions – its motor-habits (*le monde agi*) – these two together making its *Umwelt* – usually translated into English as 'environment' and into French as '*milieu*'.[2] For us the *Merkwelt* is made of information from the eyes and ears, skin, nose and tongue, and it is quite a different *Merkwelt* from that of the dog, or the tick, or the bat which have different sensory apparatuses. In buildings one can make a similar division between the aspects of the building that help us to do things – housing our practical activities – and the aspects that help us to signify things – communicating messages. Of course the division between these two realms is not a hard one: a house that signifies an elevated status might help us find a mate or to have a wider circle of friends. Nevertheless there is an aspect of the building that accommodates motor functions, and another that belongs in the realm of signification. If we go back to John Clare's cottage then it is plain that the building has no preordained place in a system of signifiance. It engenders contentment on account of the states of affairs that it brings about within it, all of which are related to the processes of subject-formation (subjectification) through family relationships, and relations with the fireplace, the tobacco pipe, the dog, the comfortable chair, the bed and so on. It is a black hole. Seen from within it does not look like a black hole, but for practical purposes it is invisible from the outside and is productive of domestic contentment through subjectification, not signifiance. By contrast the cottage of Marie-Antoinette has been facialized. It is very clearly part of a world of signification, and is in play as a sign from its outset. It is less clear that it has any part to play in any process of subjectification, as even the practical activities within are overcoded in ways that makes the practical utility of the activities into a negligible side-effect of what is going on. The erotic play is a much more important part of what is going on here than the milk-production from the dairy. Dressing up as a milkmaid – which is to say in a uniform that signifies 'milkmaid' – is overwhelmingly more important than doing any part of a milkmaid's job or than the milk-yield in this particular environment (*milieu*). Marie-Antoinette's dairy is more white wall than it is black hole, but it is an expression of something that she felt a need to act upon. There was a process of subjectification going on here, but of course it was not confined to the hamlet: the great facialized building at hand was the palace at Versailles, which was an outlandish and elaborate instrument of subjectification constructing the *persona* of the king and his relation with his court – an exemplary and unimaginably complex interplay of symbols and codes – in gilded halls, under painted vaults, twinkling chandeliers and infinite refractions and reflections. The white wall of the garden façade, with the king's bedchamber at its central point, and the black holes of the windows intimating the huge collectivity of subjectifications within. The little black hole in the innocently playful hamlet was a satellite of the huge abyss of a black hole in the palace, that would, by 1793, swallow Marie-Antoinette whole, along with the rest of the royal family, and the institution of the court.

Signifying

The façade as an architectural development is not universal, even in high-status architecture. Given Deleuze and Guattari's identification of Christ with faciality one would like to see it developing with the church, and it is certainly to be found there, in the great west façades of churches – especially those such as Salisbury cathedral or the churches in Lucca and Pisa where the west façade stretches up higher than the building behind it, so as to be able to accommodate ever more signification – more statuary, more arches, more splendour. It is there emphatically in the Romanesque church of San Miniato al Monte in Florence, where the *Wirkwelt* is accommodated in a brick and limestone structure, while the *Merkwelt* is established by attaching a thin sheet of white and coloured marbles to the external façade and the parts of the interior that belong to the world of signifiance (see below). This white wall was

Façade of San Miniato al Monte, Florence, 1090 onwards.

taken up and reflected across the city when Alberti remodelled the façade of Santa Maria Novella; and it was Alberti who attached a finely designed façade to the Rucellai palace, in the heart of the city, making a deliberately facialized domestic building. In the Rucellai palace the ragged unfinished edge of the façade makes it look particularly mask-like, and so it expresses the idea of faciality particularly clearly (see below) but it did not initiate this white wall/black hole assemblage in domestic architecture. In the ancient world domestic buildings seem to have saved their real splendours for the interiors; but Diogenes mentions showy porches on the streets of Athens, which clearly had a status-enhancing function, and as they belong in the *Merkwelt* while the private house's *Wirkwelt* activities of subjectification happened in the black hole within, one can see a domestic faciality developing, and can see Diogenes deploring it, or gleefully taking advantage of it, as he adopted the streets and porches as his own dwelling place (Diogenes, *c.* 340 BC, 41, no. 14). The role of the household and the role of the city in the production of the subject is mentioned in one of Diogenes' aphorisms: 'The road from Sparta to Athens is like the passageway in a house from the men's rooms to the women's' (Diogenes, *c.* 340 BC, 58, no. 113). Houses separated men and women, who underwent different subjectifications, and so did the differently formed citizens of Sparta and

Leon Battista Alberti, façade of the Rucellai Palace, Florence, 1452–70.

Athens. Sparta's symbolic economy was invested in the body of the warrior, which one could therefore say was facialized, while Athens' was in its monuments. The city's white wall/black hole was on the Acropolis: the white wall being the imperious monumental Parthenon, made of sparkling Pentelic marble. It had a dark interior of its own, but the real black hole was the interior of the Erechtheion, a short distance away, where the most sacred relics were kept out of the sun and not often seen; they were nevertheless celebrated and were instrumental in contructing the city's identity – Athena's robe, a folding stool made by Daedalus, who had devised the labyrinth to house the Minotaur, a stove with a palm tree flue designed by Callimachus, who devised the Corinthian capital, etc. And faciality is evident further back, not in the cave, which is a pure black hole, but in the Egyptian temple with its billboard-like pylon frontage, screening behind it, beyond a courtyard open to the sky, beyond a hypostyle hall crowded with columns, the cave-like sanctum of the most ancient chambers within. Quatremère de Quincy argued that ultimately all Egyptian architecture derived from the cave (Lavin, 1992) but it was made to perform a monumental role in its exterior – excluding the populace from within, and in doing so establishing a despot-face that had an important societal role. The Egyptian temple was a monumentalized version of the houses of the nobility, where presumably the front wall would have been almost exclusively a screen wall to separate the courtyard from the street, but in the temples it grew into a semi-autonomous mask, with gigantic incised depictions of the gods standing beside its dark orifice.

Radomes

The white wall however has a more ambiguous role in the *Merkwelt* when its role in reflecting back the images projected on to it is more apparent. Take the white walls of the geodesic domes on Menwith Hill in North Yorkshire (see page 77, and Wood, 2004). Their role is to protect the mechanisms within – sophisticated instruments that pick up messages from satellites. The carbon-fibre domes are invisible to the instruments, but they make the instruments invisible to observers from outside, so they have an enclosed inexpressive exterior, as white as Moby-Dick's hump. What we make of them depends on what we bring to them. If we feel that this is a useful installation that is protecting us from terror and evil then we will see it as a slightly surreal but on the whole a comforting presence, with these white radomes nestling gently against the side of the hill. If on the other hand we see it as an alien presence that escapes normal controls and regulation, then we can experience it as a very disturbing threat to the operation of civil society. If we go further and let it become an Ahab-like obsession then we might start to see the domes bobbing around and stalking us as we move through the countryside, like the balls that stalk Kafka's Blumfeld (Deleuze and Guattari, 1980, 169). Even when the radomes are out of sight we can sense them tracking us by way of the satellites and the navigation systems in our cars. In each of the three cases here – the benign, the threatening and the paranoid – the same building connects into a different assemblage, and produces a different experience for the visitor. The architecture-machine is differently constituted in each case, because each of these visitors brings a different set of concepts into play to make the machine that produces the affects (which is to say the experience, which is to say the architecture). The example of Menwith Hill is of interest because the white spheres escape recognition as conventional architectural signs. Normally we know how to read buildings as securely as we know how to read uniforms, and like uniforms the façades of a building can be worn as a disguise. Here we read the buildings as screens, and they allow us to project on to them our hopes or our fears, without offering explicit confirmation or denial.

RAF Menwith Hill, late twentieth century.

Deserts

Landscape reappears with another role in imaging the schizoanalytic 'subject', if a subject remains. Just as Lenz found himself in machinic engagement with his surroundings, so that there was no sense of separateness between his 'self' and the snowflakes, stars and mountain peaks, so Deleuze and Guattari describe themselves as deserts, inhabited by concepts that wander across them and move on their way, so they are being continually reconstituted and remade. 'We are deserts,' said Deleuze

> **but populated by tribes, flora and fauna. We pass our time in ordering these tribes, arranging them in other ways, getting rid of some and encouraging others to prosper. And all these clans, all these crowds, do not undermine the desert, which is our very ascesis; on the contrary they inhabit it, they pass through it, over it. In Guattari there has always been a sort of wild rodeo, in part directed against himself. The desert, experimentation on oneself, is our only identity, our single chance for all the combinations which inhabit us. (Deleuze and Parnet, 1977, 11)**

The 'individual' here is explicitly seen as multiple and political, and the process of subjectification is presented as dynamic and continuing, never as something that has reached or could reach a satisfactory conclusion. For Deleuze and Guattari living is always a process of becoming, never of contemplating an achieved 'being'. Deleuze describes Guattari as 'a man of the group, of bands or tribes, and yet he is a man alone, a desert populated by all these groups and all his friends, all his becomings' (Deleuze and Parnet, 1977, 16). There is something of the fluidity of identity of 'the man of the crowd' in Edgar Allen Poe's story, where the man participates in the identities of the various tribes and crowds that swarm through the city (Ballantyne, 2005, 204–9). He takes to an extreme, and embodies a principle in a way that only a fictional character can: the principle that we are not formed in isolation, but socially, and we are constituted by way of ideas and practices that do not originate in us but which pass through us and inhabit us and influence the things we do, occasionally perhaps consciously, but for the most part without our having any particular awareness of it happening. So the individual is seen as not so much a political entity as a politics (a micropolitics) populated and engaged, harmonious or conflicted. The image is always of lines and intensities, intersecting planes and multiple colours, atmospheres, flows – never hard dry objects, bounded forms or clear contours. And the face, this white screen/black hole assemblage, is a means of engaging with others, a way of putting into circulation certain sorts of signification that our little parliament, our *pandaemonium*, feels will help it on its way.

CHAPTER 5

City and Environment

A little order

If I try to think about what is going on in my body right now, based only on what I can feel, then I really have no idea what is happening. I have heard about some of the things that go on in bodies, and I can make myself aware of my breathing and my heartbeat, but from my own experience of my own body I can't say what it's like in my liver today, or in my bloodstream. I learnt, somewhere along the line, that I have a heart and a liver and that my blood circulates – but they're not things I would have been inferred just by sitting still and thinking about it. My conscious thought is a very small part of what I am, but somehow the part of me that consciously thinks is under the impression that the body is there mainly for its benefit. My liver might be under a different impression. So far as it is concerned, my body is the environment where it thrives. Similarly at the molecular level, the various parts of my body are made up of molecules from the food that I have taken in. Iron, calcium, hydrogen, carbon and oxygen and so on, have been rearranged by the little machines that do these things without my knowing much about it. I feel thirsty, or hungry, so I eat and drink, and then the various tissues and membranes sort out what I need and deal with it without troubling me with very much more information, and without my needing to use my conscious will to make my body digest better or in novel ways. From the 'point of view' of a molecule I am an intensification of certain sorts of molecule, a densification of molecules that are in my environment and in me; not that an individual molecule notices me as it passes through. Richard Dawkins has suggested that species make sense best as mechanisms that ensure the survival of genes (Dawkins, 1976) which is not what we would infer directly from consulting our own feelings about why we do things. At another scale: if I am in search of an intensification of humans, then I head for the city.

> The town is the correlate of the road. The town exists only as a function of circulation, and of circuits; it is a remarkable point on the circuits that create it, and which it creates. It is defined by entries and exits; something must enter it and exit from it. It imposes a frequency. It effects a polarization of matter, inert, living or human; it causes the *phylum*, the flow, to pass through specific places, along horizontal lines. It is a phenomenon of *transconsistency*, a *network*, because it is fundamentally in contact with other towns. It represents a threshold of deterritorialization, because whatever the material involved, it must be deterritorialized enough to enter the network, to submit to the polarization, to follow the circuit of urban and road recoding. The maximum deterritorialization appears in the tendency of maritime and commercial towns to separate off from the backcountry, from the countryside (Athens, Carthage, Venice). The commercial character of the town has often been emphasized, but the commerce in question is also spiritual, as in a network of monasteries or temple-complexes. Towns are circuit-points of every kind, which enter into counterpoint along horizontal lines; they effect a complete but local, town-by-town, integration. Each one constitutes a central power, but it is a power of polarization or of the environment [*milieu*], of forced coordination. That is why this kind of power has egalitarian pretensions, regardless of the form it takes: tyrannical, democratic, oligarchic, aristocratic. Town power invents the idea of the *magistrature*, which is very different from the State *civil-service sector* [*fonctionnariat*].[1] But who can say which does the greater civil violence? (Deleuze and Guatttari, 1980, 432–33, translation amended)

This is an analysis of the forces necessary to make a town live. It is, at least at the outset, uncontentious, translating into Deleuze and Guattari's characteristic vocabulary the key idea of Walter Christaller's central place theory, which has had a pervasive influence on the field of geography. Notice, though, that Deleuze and Guattari might say much the same things about what makes an individual – a schizoanalytic subject. Notice also that this town, which is always part of a network, constitutes a power of the *milieu*, a power of the environment, the *Umwelt*. If I am in the town, then it is my environment, but the town itself is between other towns, which make its environment. Any 'thing' can be described as an environment if we think of it at an appropriate scale.

Environment – *milieu*

Uexküll places particular stress on the relation between a creature and its environment. Each species of animal lives in its own world, distinct from the worlds of other creatures that have other mechanisms for sensing their world and for surviving in it. The opening section of Uexküll's book *Mondes animaux et monde humain* [*Animal Worlds and the Human World*] is entitled 'La tique et son milieu' ['The Tick and its Environment'] (Uexküll, 1965, 17) and the page is dominated by an illustration of a swollen tick inflated to many times life size. Deleuze and Guattari were fascinated by the simplicity of its world.

> **The unforgettable associated world of the Tick, defined by its gravitational energy of falling, its olfactory characteristic of perceiving sweat, and its active characteristic of latching on: the tick climbs a branch and drops on to a passing mammal it has recognized by smell, then latches on to its skin (an associated world composed of three factors, and no more). (Deleuze and Guattari, 1980, 51)**

Once it has latched on to the mammal, it gorges itself on blood, and in doing so swells to many times its previous size. Then it drop back to the ground and lays its eggs in the earth, its life-cycle complete (Uexküll, 1934, 18–19); and Uexküll asks this question: 'Is the tick a machine or a mechanic? Is it a simple object or a subject?' In order to answer it he adopts two *personae*, each of whom gives an

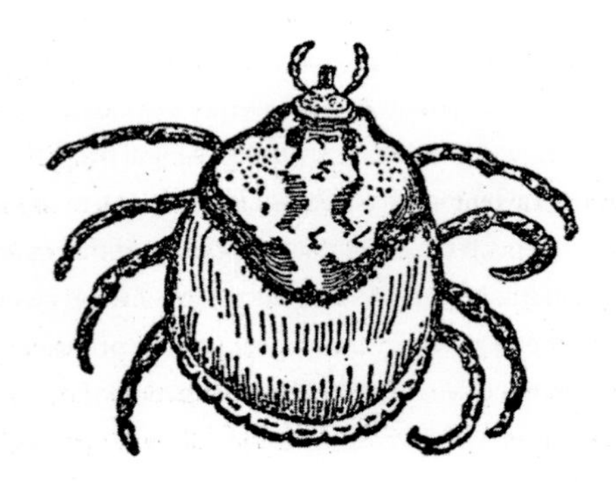

The tick.

answer mapped on a different plane. The physiologist describes the tick as a machine, saying:

> 'With the tick one can distinguish receptors (organs of sense), and effectors (organs of action), linked together through the central nervous system by an apparatus that directs. The assemblage [*ensemble*] is a machine, but nowhere is a mechanic to be seen.'
>
> 'That's just where you go wrong,' the biologist replies, 'it isn't a machine-like 'tick-body package', and there are mechanics at work everywhere.' The physiologist carries on unperturbed:
>
> 'All the tick's actions are just reflexes, and the reflex arc forms the basis of all animal-machines. It's started by a receptor, an apparatus that lets in only certain influxes from outside, like warmth or the smell of butyric acid, and rejects all the others. It's finished by a muscle that sets an effector in motion, whether it's an apparatus of walking or clinging. The sense-cells, which release the excitation of the senses, and the motor cells, which release the impulsion of movement, serve only as the linkages to conduct the waves of corporeal excitation, that are caused in the nerves by an external shock of the receptor to the muscles of the effectors. The assemblage of the reflex arc executes just one transmission of movement, like any machine. There's not the slightest trace of a subjective factor here, as if one or several mechanics were involved.'
>
> 'It happens precisely the other way about,' the biologist replies 'Everywhere we have nothing but mechanics, not parts of machines. Each individual cell in the reflex arc works not at the transmission of movement, but at the transmission of excitation. So an excitation has to be perceived by the subject and doesn't exist for objects. Every part of a machine, for example the clapper of a bell, only manages to do its machine work when it's balanced, one way and another; if you do anything else to it – apply cold, heat, acids, bases, electric current – it acts just like any other piece of metal. But we know from Jean Müller that a muscle behaves quite differently. Whatever the intervention from outside, it responds in the same way, by contracting. It changes every external intervention into the same excitation, and responds with the same impulse that causes the contraction of its cellular body.
>
> Moreover Jean Müller has shown that all the external effects that touch the optic nerves, whether they're ether waves, pressure, or electric current,

cause a luminous sensation, so our optical cells respond with the same 'perceptive character'.

Therefore we can conclude that each living cell is a mechanic who perceives and does, that consequently it has its own perceptive character, and impulsions or 'active characters'. The perception and action complexes of the assemblage of the animal subject lead thus to the collaboration of little cellular mechanics, each one of which makes use of just one perception-signal and one action-signal.' (Uexküll, 1934, 19–21)

If each of our cells is a subject, then we are indeed already quite a crowd. This touches on the same territory as Samuel Butler's life-endowed machines (see Chapter 2). Under one description the creatures look like mechanisms, and under the other the machines look as though they are alive. Either description sits uncomfortably with common sense, if it is adopted to the exclusion of the other. One description would see the living organism's aptitudes and behaviours as emergent properties deriving from a multiplicity of complex machine-like interactions, while the other would see a pervasive vitalism even in inorganic matter. In the common-sense world we use one of these descriptions for creatures and the other for uncommunicative mechanisms, but there is no clearly defined threshold point in things that separates one from the other, just a cultural habit that makes us draw a line and put things in quite separate categories, and feel uneasy, as if there is some trickery going on, when Butler's rigorous logic points out the confusion. Deleuze and Guattari's work makes use of both of these descriptions, without acknowledging any common-sense thresholds. If the desiring machines at the opening of Anti-Oedipus are a clear evocation of one, then their treatment of concepts as living entities is evocative of the other. 'Leroi-Gourhan has gone the farthest toward a technological vitalism taking biological evolution in general as the model for technical evolution: a Universal Tendency' (Deleuze and Guattari, 1980, 407 referencing Leroi-Gourhan, 1945). Guattari identified three ecological 'registers' – the environment, social relations, and human subjectivity (Guattari, 1989, 28) – and in each of them comparable processes are operating, each of them at a different scale, each establishing the various *milieux* in which the various creatures, machines or concepts take shape. 'There is an ecology of bad ideas,' said Gregory Bateson, 'just as there is an ecology of weeds', and Guattari used this good idea as the epigraph of *The Three Ecologies* (Guattari. 1989, 27).[2]

The ideas need a *milieu*, made up of other ideas and practices, and a given *milieu* will allow some ideas to flourish while others do not stand a chance. The particular *milieu* for ideas that Guattari sees threatening diversity is that of 'Integrated World Capitalism', which we would now normally call 'globalization'. It has a tendency to make us all want the same things, wherever we are in the world, and whatever our cultural differences would have been in the past. This homogenizing of the subject by the mass media presents the same sort of danger as threats to biodiversity. We are educated all to swoon at the sight of the same film stars, to order the same carbonated drinks and wear the same perfumes. At one level it is pleasurable and innocuous, and too anodyne to look as if it could possibly do any harm. On the other hand whole species of ideas and cultures of behaviour are eliminated from the planet, never to be seen again, driven out for want of attention because we were thinking about football scores, or celebrity gossip.

It makes no sense to think that an organism stands a chance of survival independently of the survival of its *milieu*; the *milieu* is a precondition for the organism's development.

The animal and its *milieu* are intimately linked, as Uexküll showed particularly clearly, and Bateson argued that they were so inseparable that they should be thought of as 'the unit of survival'. It makes no sense to think that an organism stands a chance of survival independently of the survival of its *milieu*; the *milieu* is a precondition for the organism's development; and if Deleuze and Guattari tell us always to begin in the middle, their word for 'middle', *milieu*, also means 'environment'. If we frame the organism-plus-*milieu as* a 'unit' as Bateson suggests, then it is impossible to define it neatly as having a clear form, or limit. If we can say that the tick 'is' the neat little tick-body-package that is shown in the figure, then we might feel that it has been protected if we have it in a glass jar, while its environment has been changed so that it can no longer survive there. Perhaps the mammals who crossed the site have been diverted, or hunted down, and despite the fact that no tick has been harmed, the tick population might be wiped out. However it is relatively easy to isolate and describe the tick, and to illustrate what one looks like. It is much more difficult to describe and specify this organism-plus-*milieu* unit of survival which is 'formless' if we try to define it in traditional Euclidian or Cartesian terms – thinking of it in terms of squares and circles, or as lines on graphs. It would have to be defined relationally, by explaining how one part interacts with the others, and using those networks of relationships – the politics – to say what was happening there. It is notorious that the sciences are most comfortable in dealing with medium-sized dry objects – we have the most developed ways of talking about things we can see and touch, and which stay the same from one sighting of them to the next. We have to understand miniscule and vast things by making analogies with things that are closer to the range of things that we can perceive directly, and the study of fluids and flows is in its infancy compared with the study of straight lines and cubes, which our mathematics finds so much easier to define, but which are much more exceptional in our experience of nature. '*Formless*,' said Bataille, 'is a term that serves to bring things down in the world. . . . What it designates has no rights in any sense and gets itself squashed everywhere, like a spider or an earthworm. In fact for academics to be happy, the universe would have to take shape. All of philosophy has no other goal: it is a matter of giving a frock coat to what is a mathematical frock coat. On the other hand, affirming that the universe resembles nothing and is only *formless* amounts to saying that the universe is something like a spider or spit' (Bataille, in Ballantyne, 2005, 5). Things that have form have a status and respectability that the formless does not have. With the formless one cannot even be sure that one is dealing with 'things'. *Milieux* – environments – are

. . . form corresponds to what the man in command has thought to himself, and must express in a positive manner when he gives his orders.

formless. Deleuze and Guattari's treatment of things turns them into mobile constellations of points, singularities, forces, intersecting planes or lines in their *milieux*. 'Form' takes a holiday. They draw attention to the social and political dimension of form, citing with approval Gilbert Simondon saying 'form corresponds to what the man in command has thought to himself, and must express in a positive manner when he gives his orders: form is thus of the order of the expressible' (Simondon in Deleuze and Guattari, 1980, 555, n. 33). Towns, environments and subjects ('individuals', or 'dividuals' as Deleuze and Guattari sometimes call them in recognition of their divisibility) are presented as correlates of relation, the product of networks and flows.

Disconnecting

The state is presented as something quite different, having a tendency to disconnect from the wider networks:

> The State indeed proceeds otherwise: it is a phenomenon of *intra-consistency*. It makes points *resonate* together, points that are not necessarily already town-poles but very diverse points of order, geographic, ethnic, linguistic, moral, economic, technological particularities. It makes the town resonate with the countryside. It operates by stratification; in other words, it forms a vertical, hierarchized aggregate that spans the horizontal lines in a dimension of depth. In retaining given elements, it necessarily cuts off their relations with other elements, which become exterior, it inhibits, slows down, or controls those relations; if the State has a circuit of its own, it is an internal circuit dependent primarily upon resonance, it is a zone of recurrence that isolates itself from the remainder of the network, even if in order to do so it must exert even stricter controls over its relations with that remainder. The question is not to find out whether what is retained is natural or artifical (boundaries), because in any event there is deterritorialization. But in this case deterritorialization is the result of the territory itself being taken as an object, as a material to stratify, to make resonate. Thus the central power of the State is hierarchical, and constitutes a civil-service sector; the centre is not in the *milieu*, but on top, because the only way it can recombine what it isolates is through subordination. Of course there is a multiplicity of States no less than of towns, but it is not the same type of multiplicity: there are as many States as there are vertical cross sections in a dimension of depth, each separated from the others, whereas the town is inseparable from the horizontal network of towns. Each State is a global (not local) integration, a redundancy of resonance (not of frequency), an operation of the stratification of the territory (not of the polarization of the *milieu*). (Deleuze and Guattari, 1980, 433)

So it is in principle a straightforward business to draw a map of nations on a sheet of paper, because – except in disputed territories – there is a boundary that cuts the state off from the surrounding states. The boundaries might be redrawn from time to time, but in principle the centrally determined laws operate up to the state's limit and not beyond. The crucial point here is that the 'centre' where the decisions are taken is not in the *milieu*, but above it, outside it, on another stratum. So this description of the state and its organization correlates with the 'hylomorphic' conception of form – derived from Aristotle's ideas, and here importantly to be contrasted with the idea of 'emergent' form, or *immanence*. The substance of the state is formed by a power that acts from a higher stratum than the substance. The town-networks are immanent in their *milieux*. The state has form, the town is formless. Of course towns can have order imposed upon them from a higher stratum, but that is not what makes them work, and it is no way to understand urban design. Towns make the *milieu* for individual buildings, and one needs to understand the interdependence of building and *milieu* if one is to design a successful building – a building that sustains life, and that becomes a thriving organism. The factors that make people, or buildings or towns live and work and thrive are formless and need to be understood, but they operate in *milieux* that are cut across by various apparatuses that act like the state, tending to separate a part of the network from its wider surroundings. For example, the ownership of land is regulated in ways that are like the setting up of state boundaries, and in some respects my state-defined legal responsibility stops at the edge of my land. Certainly if I am inclined to act as a non-transgressing citizen then that is going to be where my building has to stop. Traditionally architecture has been preoccupied with form, for example in Le Corbusier's definition: 'Architecture is the masterly, correct and magnificent play of masses brought together in light' (Le Corbusier, 1923, 29). This is delivered from a higher stratum: 'magnificent' is clearly above the *milieu*, and 'masterly' and 'correct' behaviours conform to a pattern determined from above. And we have learnt to see form (here 'masses') as 'what the man in command has thought to himself,' and has been able to express. So Le Corbusier's definition of architecture belongs entirely to the mindset of the state, and we can enlist him to the service of the *fonctionnariat* and have him design buildings as limited well-defined object-parcels that tend to separate themselves from their surroundings. The cult of pure form, of beautiful shapes that enchant us with their other-worldly promise of an unencumbered life, is the staple of the glossy architectural magazines. The immanent order in the life played out in buildings remains undiscovered in these images, which prefer to show how closely one can aspire to live in surroundings that have geometric definition or well-defined pictorial qualities. Immanent order might emerge at a domestic scale if unselfconscious housekeeping routines were the exclusive determinant in forming the house, but, if we can, we usually try to shape things so as to lay claim to status of one sort or another, for example by making the house in some way look like a house. Most of us, most of the time, have an idea of what a house looks like. Our sense of form derives not only from the emergent properties of the *milieu*, but also from the regimes of signs that surround us, and that we deploy. Where human buildings are concerned, emergent form is more evident at the scale of the city, where the individual buildings might be self-conscious but where the wider picture is often left to take care of itself. Friedrich Engels described this happening in nineteenth-century Manchester, when there was an astonishing boom, and over the course of only a few decades it was transformed from a village into a metropolis. The surprising thing here was that despite the evident free-for-all, a clear order did emerge.

> **The town itself is peculiarly built, so that someone can live in it for years and travel into it and out of it daily without ever coming into contact with a working-class quarter or even with workers – so long, that is to say, as one confines himself to his business affairs or to strolling about for pleasure. This comes about mainly in the circumstances that through an unconscious, tacit agreement as much as through conscious, explicit intention the working-class districts are most sharply separated from the parts of the city reserved for the middle class [. . .] Manchester's monied aristocracy can now travel from their houses to their places of business in the centre of town by the shortest routes, which run right through the working-class districts, without even noticing how close they are to the most squalid misery which lies immediately about them on both sides of the road. This is because the main streets which run from the Exchange in all directions out of the city are occupied almost uninterruptedly on both sides by shops, which are kept by members of the middle and lower-middle classes. In their own interests these**

shopkeepers should keep up their shops in an outward appearance of cleanliness and respectability; and in fact they do so [. . .] Those shops which are situated in the commercial quarter or in the vicinity of the middle-class districts are more elegant than those which serve to cover the workers' grimy cottages. Nevertheless, even these latter adequately serve the purpose of hiding from the eyes of the wealthy gentlemen and ladies with strong stomachs and weak nerves the misery and squalor that form the completing counterpart, the indivisible complement, of their riches and luxury. I know perfectly well that this deceitful manner of building is more or less common to all big cities [. . .] I have never elsewhere seen a concealment of such fine sensibility of every thing that might offend the eyes and nerves of the middle classes. And yet it is precisely Manchester that has been built less according to a plan and less within the limitations of official regulations – and indeed more through accident – than any other town. (Engels, 1845, 84–6)[3]

Engels explains how the pattern of the city is generated not by the imposition of form from a higher level, but by decisions taken within the *milieu,* especially but not exclusively by the small shopkeepers. He calls the manner of building 'deceitful', and in doing so places himself on a higher stratum, because from within the *milieu* that is not the way it looks. If anybody does notice what is going on (and Engels leads us to believe that they don't) then what they would tell us, if we asked, would be that appropriate judgements were being made about what type of building belonged in each type of place. It would look like a matter of decorum, not deceit or hypocrisy. The *milieu* in which one lives is inhabited not only by other people with whom one interacts, but also by animals, vegetation, flakes of snow and mountain peaks, and by ideas – including on occasion ideas about how to deal with buildings – that are part of our ecology. So here in the Manchester that Engels saw, but of which he was not altogether a part, the ideas about architectural decorum were widely shared among the people who had the means to act upon them. They did not need to take an overview of the whole, but only to see where it would make sense to open up shop, and how to run the place in order to make a decent living. It involves no radical thinking, but a pervasive common sense that within the *milieu* seems to remain unchallenged. From outside and above it looks deceitful and as though there has been some sort of brainwashing. From within it looks

From outside and above it looks deceitful and as though there has been some sort of brainwashing. From within it looks as if things are going swimmingly.

as if things are going swimmingly. The city – *this* city, at least in this account of it – is a self-organizing system, whose order is immanent. In the same way the account of the schizo-analytic subject – the individual person – at the beginning of *Anti-Oedipus*, is a self-organizing system which under one description has a unified will and a personal name, but under another description is a teeming swarm of desiring-machines that have no way of forming a view of the whole. Just as a person can be coerced into the adoption of inflexible social roles, such as those offered by the 'holy family' – the nuclear family unit, which is presented as useful to capitalism and Oedipalizing in its effects – so can a city be given an appearance of correctness and magnificence that may not help it to live. The imposition of 'form' might give a city the appearance of respectability and high status, but if it does not mesh with the networks that generate the city's life then it will be left with deserted boulevards and windswept plazas that might look good in photographs but which will not help the place to flourish. It would be far better to find oneself in an unselfconscious city like Engels's Manchester that does what it has to do without making a claim to cultural status for itself. There were of course grave problems with Manchester, and many people lived in abysmal conditions, but there was no doubting the city's overall vitality. The surprise was the apparent clarity of its organization, given the lack of any centralized planning control. In order to generate another city like Manchester one would not specify a form, but would put in place the conditions: a world-beating commercial operation that has need of a large workforce (much of it with limited skill, and therefore poorly paid). The rest more or less follows as a consequence. The people who set things in motion become very rich, and although they are a small class of people they have the money to dispense to see that their desires are acted upon. The people on low wages have access to a different range of things, which cost less and are more widely available. The needs of every level of society are met by service-providers who are dependent on the central commercial operations, but at one or more removes. It is this middling class of the *milieu* – which might have an official wing in a *magistrature* – that seems to be critical in determining the decorum of the place, selecting the places where the shops will be set up, and making the best of their façades. So long as this lower-middle class has a shared sense of propriety, and the other classes do not overwhelm it, then its pervasive sense of order can prevail without reference to centralized control mechanisms. It is Middle England, and in Manchester especially the small shopkeepers, who seem to have the decisive impact on the city. Heroic architects design one-off oddities in the city, but its fabric is apparently unselfconscious emergent 'design', for which an individual can make no claim to authorship. It is the outcome of thousands of local decisions, much as we see in ant colonies that build themselves anthills, and slime-mould communities that seem to solve the problem of finding the shortest route across a labyrinth (Johnson, 2001).

The specification of the relations that produce these effects could be described as political, or else as topological – topology being 'rubber sheet geometry' which sees things as equivalent if one can be stretched or folded into the other (so all closed bounded shapes are equivalent, whether they be circular, square or irregular; likewise a sphere and a cube would be like one another, but different from a torus (doughnut shape) which has a hole in it whereas the others do not. Deleuze and Guattari draw attention to the way that matter and form are not inseparable, that materials have '*singularities or haecceities* that are already like implicit forms that are topological rather than geometrical, and that combine with processes of deformation: for example, the variable undulations and torsions of the fibres guiding the operation of splitting wood' (Deleuze and Guattari, 1980, 408). Notice here the 'implicit' forms – virtual forms – which are 'folded' into the material. Deleuze's study of Leibniz was entitled *Le Pli* (1988, *The Fold*) and this kind of fold is very much more his subject matter than any literal foldings of pieces of paper of sheet metal. The fold, *pli*, is in action in such words as im*ply*, im*pli*cit, multi*ply*, du*pli*cate, re*pli*cate – they are all 'folding' words. It is not clear that these implicit forms could be actualized, but they have an influence on the outcome. In the example of timber, the traditional craftsman would be very familiar with the effects of the grain of the wood and would know how to use it to his advantage. However he would not normally be working to express it, or to allow it to find expression. There are examples when

Ladder at Epoisses, Côte d'Or, date uncertain.

it might happen. A piece of timber that follows the grain of the wood is stronger than one of the same dimension that has been sawn geometrically straight. A good strong walking stick could be made by searching out a good straightish growth from a hedgerow, cutting it to length and working it up. Although it would be strong it would not look sophisticated, and would carry the traits of rusticity – the low status connotations of Bataille's 'formless'. A sophisticated urban walking stick would be straight, even if that meant that it was not so strong. In the *pigeonnier* of the chateau at Epoisses, there is a remarkable ladder which gives access to all the nesting places in the wall of this great cylindrical room. The ladder is mounted on a spindle in the centre of the room, and it can be turned to reach any part of the circumference. The steps are held in place by two stringers, which match each other exactly because for each section they were split apart from one piece of wood, and these timbers were spliced together to make a single run of surprising slenderness (see p. 93). The sacrifice, if sacrifice it be, was that the ladder could not be straight – it had to follow the line of the tree's growth, the grain of the wood. It is a beautiful and skilfully wrought thing, and it is technically sophisticated, but it does not conform to the geometric idea of good form, and so there is nothing remotely like it in the polite rooms of the chateau, only here where the pigeons lived, and where the gamekeeper would visit. Before the mid-nineteeth century, when Viollet-Le-Duc and some others started to promote the idea that materials should have a say in the form they were asked to make, whenever the grain of the materials came through, the character was construed as having low status, verging on the formless, as if the craftsman did not have the skill to overcome the material and bend it to the inflexible will to geometric form. The character of the materials in a simple woodsman's cottage, for example, is very evident, and the shepherd who managed the hefted sheep would have lived in a dwelling that was an intensification of the stones from the fields, just as he himself would have been an intensification of elements from the crops and flocks.

Emergent form

By the time the builder starts on site, the architect is normally acting from another stratum, issuing instructions about a form that has already been determined in the studio, so that a builder can act on the instructions in order to actualize the form. It is far from being a practical impossibility, but in a commercial operation, such as most architectural practices feel the need to be, one is rarely in the position of being able to experiment directly with unformed materials, wresting them directly from the ground. Nevertheless it is a position to which one can aspire, and when fundamental innovations can be made. The micropolitics of materials at a molecular level translates into their properties at the level where we can perceive and make use of them: the different molecular structures of limestone and slate, for example, mean that the different stones make different characteristic shapes, and we have found ways in which these characteristic stone-shapes can be useful to us in making walls and roofs. If we turn our attention to what is practically available to us on a given site, then for readers of this book the range of building materials certainly includes things from builders' suppliers – concrete blocks of standardized dimensions, timber that has been through sawmills and cut to predictable regular sizes, steel nails, screws and other jointing methods, and any number of sheet materials from plate glass to insulation boards and medium density fibreboard. These are the materials that a builder would normally be expecting to work with, and the forms that they generate are not identical with those generated by field stones and forest thinnings. As soon as one steps into a normal commercial contractual relationship to produce a building, where speed of delivery and reliability of result are paramount, then experimental thinking tends to be excluded from consideration, and common sense takes over, producing much the same results as before. One has to fight against this tendency if one is to produce innovation. In normal circumstances, the work of hylomorphism is all but done before the materials come into our hands. Moreover the decision to devote time and energy to resisting it on the plane of materials could well reduce the quick and effective production of something that would be radical and effective on the social plane. For example the politically effective shelters for homeless people that appeared in Paris in the winter of 2006 were off-the-peg tents bearing simply hand-painted marks (the group called itself 'the children of Don Quixote' – after the would-be knight errant, whose romantic self-delusion and hopelessness are ironically adopted so as to point out the hopeless idealism of a group that would hope to house the homeless) (see p. 96). It would have been absurd and counterproductive had money and effort been expended on making brilliant new tent designs for this purpose. The tent designs do however make

Tents provided by Les enfants de Don Quichotte, along the Canal Saint Martin in Paris, December 2006.

ingenious and effective use of the materials from which they are made. A steel 'spring' is held in place and is held taut by the tent's fabric that also makes the protective shelter. So the politics going on in the steel molecules' *milieu*, is brought into contact with and held in an agonistic relationship with the *milieu* of the molecules in the canvas, which nevertheless remain unmixed and untroubled by one another (there is no chemical reaction between the fabric and the steel). The *milieu* within the tent – its internal climate significantly more habitable that the external one – is established by the properties of the fabric, which resists the passage of air and water, and is warmed by the body of its inhabitant. The political *milieu*, in which the police, ideas of human rights and ideas about decorum in important public places can and do interact with one another becomes critical at this point of intensification where the planes intersect.

Form and frame

Lose sight of form as an intention – immerse yourself in the politics of the molecules, the lives, the affects in the various *milieux*. Propose something in the *milieu*. The *milieu* has as many dimensions as one takes the trouble to notice, and one weighs their importance by projecting them on to an ethical plane. Architects sometimes like to make the claim that architecture is autonomous, but to make such a claim is merely to deny the legitimacy of some of the multiplicity of planes, which nevertheless remain real even if we do not allow ourselves to talk about them. Finding a form for a building has a parallel in finding form in oneself. One fixes a limit – a frame. I decide that I am the kind of person who does some things and who would never do some other things; and then sometimes at formative moments, I realize that I have to revise the idea and that I'm not quite (or not only) the person I had thought I was. And how much more true this is of other people, where we make our surmises based on much less evidence. We get to know characters in novels by hearing about what they do, and sometimes by hearing what they think about those things. With my closest friends I can 'be myself', but in various ways I would be more guarded and more formal in dealing with recent acquaintances. There is a corresponding range of decorums in buildings, from the apparently informal operations in personal private space (even though on closer inspection we will find them to have been culturally constructed) to the places where the formality is more evident – where uniforms are worn, and where 'formless' behaviour looks inappropriate or unprofessional and excludes one from the 'game'. The debating chamber of a town hall is asked to frame an activity that has a significance for the whole local community, and it should have found a way of signifying a status that is higher than that of, say, my kitchen. Architecture is good at monumentalizing the institutions that a society values, finding ways to frame the activities that are seen to be valuable in one way or another. However, Deleuze and Guattari's thought tends to promote un-monumental aspects of life, preferring fluidity and creativity ('becoming') to establishing any sort of fixity. Their thought is a challenge for any architects who choose to engage with it, as its volatility is at odds with the profession's traditional preoccupation with form. With Deleuze and Guattari one leaves behind the well-defined forms of solid objects, for a description of relations between unformed elements (longitude) and sets of affects (latitude), in order to construct a map of a body – and a 'body' here can be any entity at all, clear or vague, from an idea to a whole world, including along the way of course such bodies as people, buildings and their environments. 'The longitudes and latitudes together constitute Nature, the plane of immanence or consistency, which is always variable and is constantly being altered, composed and recomposed, by individuals and collectivities' (Deleuze, 1970, 127–8). The descriptions proposed by Deleuze and Guattari, then, are descriptions of virtualities, which can be taken up and actualized, composed and recomposed. So rather than offering a set of instructions upon which one could immediately act, their programme involves unmaking everything that one knows, and then showing how, with just a little order, a world takes shape and dissolves. The environments of ideas and of creatures shape the invention or development of those entities, just as environments of ideas, materials and politics, shape buildings and the environments within and around them that in turn shape us. The 'plane of immanence' is an environment in which the various forces act, and which produces a body when a little order resonates through it – a refrain, a concept – and the body once produced is a part of the environment and an influence on its future becoming.

Look at the mountain, once it was fire.

At certain moments, small changes in the initial conditions or in the balance between the forces that act can produce results that are very different. The ecology of ideas and the flow of capital that produces our dwellings and our cities, that produces a glittering citadel of mirrored towers here, and a flowering of neocolonial suburbs there, can be accepted uncritically or can be opposed, but it helps to understand that the surface appearances are produced by little mechanisms driving greater ones without immediate reference to a wider picture. This is the level at which Deleuze and Guatttari give us an apparatus with which to make an analysis of what is going on, and to see how everything is connected without every part being conscious of having wider connections. In a way these processes are already finding expression in everything around us, including the things we think and do. However another challenge for the architect as an artist would be to find ways to make us feel the reality of these processes, as Cézanne made the landscape speak of its formation: 'Look at the mountain, once it was fire' (Cézanne, quoted by Deleuze, 1985, 328, n. 59). Buildings are inescapably expressions of the great forces that shape them, whatever one might try to do about it, and someone looking back at them would be able to see immediately when they were built and maybe understand why, and they would be able to infer these things whatever the intention of the building's designer.

. . . aim to give voice to the song of the Earth, to show by way of some glimpse of chaos how there were other possibilities.

The processes involved in globalization, for example, would escape an individual's control, but could be importantly at work in producing the building and shaping it. But other aspects of the programme might be expressible given the right guidance. One could aim to give voice to the song of the Earth, to show by way of some glimpse of chaos how there were other possibilities, and how the building that emerged was actualized from the chaos of virtualities. A great monument would restructure the world, based on a little order taking a hold in its chaos, and working its way through into the form of the building and into the kinds of lives that can be led by the people who come into contact with it, making a framework for those lives, or part of a framework. A building is formed in a *milieu*, but it also has a *milieu* within and around it, where new concepts and new ways of living can be shaped. The formative territorializations here, though, are things that Deleuze and Guattari themselves would be trying to go beyond, to mobilize and deterritorialize, so that, having developed to a certain extent, one opens up to chaos, makes oneself receptive to what one finds there, steps outside the structured world of habits and common sense, and sees what happens. Just as the subject, oneself, is more clearly and comfortably a self when it is unselfconscious – playing backgammon – so the object can reach its best form when the 'designer' is dispersed into a multiplicity, that has its minds on other things.

. . . one opens up to chaos, makes oneself receptive to what one finds there, steps outside the structured world of habits and common sense, and sees what happens.

Further Reading

What should you read next? Of course you must read Deleuze and Guattari themselves. They wrote four books together, and many more separately. This book has discussed a few of the many ideas in the two volumes of *Capitalism and Schizophrenia*, and *What is Philosophy?* The bibliography that follows is very far from complete, but if you search on the web you will quickly locate an up-to-date one (I do not cite a particular one, as web addresses can change). There are now several helpful introductions – more to Deleuze than to Deleuze and Guattari, because they are mostly written by philosophers who are more interested in Deleuze's ideas. They tend to assume that their readers have a grounding in philosophy or that will be interested in the resolution of philosophical questions that are raised in Deleuze's work. In saying what I go on to say below, I make the assumption that the readers of this book will have an interest in architectural issues and a grounding in architectural rather than philosophical literature.

When one approaches an unfamiliar and ill-defined body of work, there is always a feeling that one does not know enough to understand it properly. This is certainly true of Deleuze and Guattari's work, wherever one begins with it. There is some encouragement to be had in Deleuze and Guatttari's sense that what we take away from the book will not be what they put into it. The point is not to come away thinking Deleuze and Guattari's thoughts, but rather to come into contact with ideas that derail our usual habits of thought and allow us to come away energized, thinking our own thoughts that might be quite unlike any ideas that Deleuze and Guattari might have had on our behalf. I think that this is something that architects are already inclined to do, whether or not the author has given permission for it. Creative misunderstanding, or misprision, is legitimate behaviour in the Deleuze-and-Guattari-world. It would be unsatisfying to do something in Deleuze and Guattari's name that betrayed them, turned them into sages whose wisdom would be reverentially unpacked and codified into an orthodoxy. Of course that is the tendency of scholarship in general and of a book such as this one, which purports to offer a way into the Deleuze-and-Guattari-world. It is more accurately a way into the Deleuze-Guattari-and-Ballantyne-world, which in some respects is quite different from the worlds produced by other readers, who have connected with different aspects of Deleuze and Guattari's work.

The first directive for 'further reading' must be to read Deleuze and Guattari, if only to understand what all the fuss is about, and why there is a demand for commentaries that make their work more approachable. The two volumes of *Capitalism and Schizophrenia* are a good place to start, as they are conceived as prompts for creativity rather than as chains of carefully presented reasoning. There is close reasoning here, but it is not set out in the way that one would expect in a philosophical text. Substantial arguments are often 'taken as read', perhaps because they are made in another context either by Deleuze or Guattari, or by someone else altogether, whose work is mentioned in passing. Their method is to give concrete examples of a concept in action, and these examples are often arresting. Part of the method pursued in my own text above is to start with one of these examples, or the concept that it is presented as exemplifying, and to see where it leads – back into the *milieu* from which Deleuze and Guattari wrested it, forward into the connections it makes in my own experience. There is no substitute for an encounter with *Anti-Oedipus* and *A Thousand Plateaus*, but equally they do not work by brief exposure and a quick read-through. Expect to read and be baffled, and from time to time there is an idea that connects and generates excitement, that makes one feel that the effort has been worthwhile. Some ideas lodge in the mind, and linger there, and come to seem significant even though they didn't seem to be at first. It is with continual revisiting, especially after having read something related from elsewhere in Deleuze and Guattari's corpus of work, or from the texts that they discuss, that the ideas settle into place as a connected web. It does not happen all at once.

Some of the connected thinking that is drawn upon in *Anti-Oedipus* and *A Thousand Plateaus* was worked out in more detail in two of Deleuze's earlier books: *Difference and Repetition* (1968) and *The Logic of Sense* (1969). They are densely written and do not have the playful allusive qualities of *Anti-Oedipus* and

A Thousand Plateaus, so this side of things is best approached by way of Todd May's *Gilles Deleuze: An Introduction* (2005) which is very clear-headed and illuminating, starting out from the question 'how might one live?' which is a question that one would want to be pervasive in an architect's thinking. Reidar Due's *Deleuze* (2007) could also be helpful in giving an understanding of some of the arguments that are implicit in Deleuze and Guattari's collaborative work; and so also would be Claire Colebrook's Deleuze books (2002, 2006) and Jean-Jacques Lecercle's study of the literature of *délire, Philosophy Through the Looking Glass* (1985) which includes material on Deleuze and Guattari. These books are not about architecture, but would help with a general orientation in Deleuze and Guattari's thought. The same could be said for the essays and interviews collected in two volumes that were published after Deleuze's death: *Desert Islands* (2002) and *Two Regimes of Madness* (2003). These include journalistic pieces that helped to explain and promote the books when they were published, and are less 'technical' in character than the books themselves, so they make a considerate introduction from Deleuze himself. The same could be said, but to a lesser degree, for the other collections of essays: *Negotiations* (1990) and *Essays Critical and Clinical* (1993). The book with Claire Parnet, *Dialogues* (1977) is ostensibly an introduction to Deleuze's thought, and is an overview, but it would be puzzling as a first introduction.

The work specifically on architecture that carries the purest Deleuzian pedigree is Bernard Cache's *Earth Moves: The Furnishing of Territories* (1995). Cache participated in Deleuze's celebrated seminars at Vincennes and Deleuze referred to Cache's then-unpublished work. 'Inspired by geography, architecture, and the decorative arts, in my view', said Deleuze, 'this book seems essential for any theory of the fold' (Deleuze, 1988, 144). Deleuze associated Leibniz's mathematics – the infinitesimal calculus, which deals with rates of change – with the infinite folds of baroque ornament, and then with the infinite regress of fractal geometries. Various ideas of the fold, some of them Deleuzian others not, were brought together by Greg Lynn in *Folding in Architecture* (1993). Lynn by way of digital technologies, has pursued the development of complex formal invention in his architecture projects (Lynn, 1998a, 1998b and 2006) and digital handling of emergent form is explored under the name of 'morphogenetic' design (Hensel, 2004 and 2006). John Rajchman's approach in the essays collected in his books *Constructions* (1998) and *The Deleuze Connections* (2000) are explorations of Deleuze's concepts that draw on a cultural background in architecture, so although they are sophisticated works they are approachable by someone who has experience in the architecture-world who is not intimidated by encountering the names of philosophers on the page. Individual essays about architecture that deserve mention include Paul André Harris on the Watts Towers (in Buchanan, 2005) and Ian Buchanan on the Bonaventure Hotel, also in Los Angeles (in Buchanan, 2000, 143–69). This last is also collected in my own *Architecture Theory* (2005, 272–300), which I of course recommend as further reading because it takes up some of the ideas presented in this book and further opens up the Deleuze-Guattari-and-Ballantyne-world. The experimentalist side of their thinking is linked there with the American Pragmatist tradition, and with architecture.

Manuel Delanda has written extensively and compellingly about Deleuze in connection with things that interest architects, such as his conceptualization of physics (2002) which has a bearing on building materials and how one would think of using them. Such an intuition generates the thinking in Reiser and Umemoto's *Atlas of Novel Tectonics* (2006). Delanda's *A Thousand Years of Nonlinear History* (1997) is a breathtaking *tour-de-force* that is grounded in Deleuze and Guattari, taking up aspects of their thought that generate narratives working at scales that are utterly unlike the scales of human experience, from the vast and sustained scales of geology to the molecular scale of genes, and his project is pursued by other means in *A New Philosophy of Society* (2006). Guattari's early works connecting social, clinical and political ideas are represented in *Molecular Revolution* (1984) but architects will find him more approachable by way of his last books: *The Three Ecologies* (1989) and *Chaosmosis* (1992), which set out the thinking that underpinned his activism in green politics.

Deleuze and Guattari connect with so many ideas and issues, that the list of possible further readings proliferates rapidly and uncontainably. They are dispersed. They are everywhere. When an idea takes hold, one should run with it, and it will zigzag here and there to arrive at unanticipated points, which will never be conclusions.

Notes

1 Who?

1 Deleuze's nails were long, and the subject of fretful remarks by Michel Cressole. See 'Letter to a Harsh Critic' in Deleuze, 1995, 5.

2 Victor Delbos (1862–1916) wrote two books about Spinoza: *Le Problème moral dans la philosophie de Spinoza et dans l'histoire du spinozisme* (Paris: Alcan, 1893) which Deleuze describes as 'much more important than the academic work by the same author' *Le Spinozisme* (Paris: Vrin, 1950).

3 This is from a transcription of an interview chaired by Maurice Nadeau that includes contributions from Guattari and others, which was published originally in 1972 when *Anti-Oedipus* was newly published. It is collected and translated in *Desert Islands* (Deleuze, 2002) under the title – more of a headline – 'Deleuze and Guattari Fight Back . . .'. Deleuze and Guattari's working methods can be examined more closely now that those of their annotated papers that were left in Guattari's possession have now been published (Guattari, 2005).

2 Machines

1 The entire text of Samuel Butler, *The Book of the Machines* is included in Ballantyne, 2005, 126–43.

2 The translation back from the French '*bourdon*' turns Butler's own (now archaic) 'humble bee' into a modern 'bumblebee'. There is an erotic encounter between an orchid and a *bourdon* (but not a wasp – *guêpe*) in Proust. It occurs in Part 1, of *Sodom and Gomorragh,* which throughout uses erotic images of plant stamens and insects brushing against them. However, Deleuze did not dwell on it in his book about Proust (1964). See also 'the marriage of bumblebee and snapdragon' in Deleuze and Guattari, 1994, 185. The editor's essay that introduces *The Anti-Oedipus Papers,*

entitled 'Encounter Between a Wasp and an Orchid', turns Deleuze and Guattari into these protagonists (Stéphane Nadaud, in Guattari, 2005).

3 *Adaptation* (2002) directed by Spike Jonze, written by Charlie Kaufman, is an elaborately fictional story about the adaptation for Hollywood of a literary non-fiction book, *The Orchid Thief* (Orlean,1998).

4 On Western agriculture of grain plants and Eastern horticulture of tubers, the opposition between sowing if seeds and replanting of offshoots, and the contrast to animal raising, see Haudricourt 1962 and 1964. Maize and rice are no exception: they are cereals 'adopted at a late date by tuber cultivators' and were treated in a similar fashion, it is probable that rice 'first appeared as a weed in taro ditches' [Deleuze and Guattari's note, 1980, 520].

5 Arguably the concept is already in play in *Anti-Oedipus*, but it was localized in the rhizome by 1976, when Deleuze and Guattari published a book of that name, which would be reworked as the introductory chapter in *A Thousand Plateaus*.

6 The date is given enhanced significance by being included in the title of a chapter in *A Thousand Plateaus*, 'November 28, 1947: How Do You make Yourself a Body Without Organs?' (Deleuze and Guattari, 1980, 149). Artaud's 'body without organs' had made an earlier appearance in Deleuze's work, before his published collaborations with Guattari (1969, 88).

7 Eileen Gray, quoted by Peter Adam, (Adam, 2000, 309).

3 House

1 Gregory Bateson (1949) ' Bali: the Value System of a Steady State', in *Architecture Theory* (Ballantyne, 2005, 74–87). Bateson cited by Deleuze and Guattari, 1980, 158.

2 For *Scenopoetes dentirostris* behaviour Deleuze and Guattari reference Marshall (1954) and Gilliard (1969). The same remarks are to be found in Deleuze and Guattari, 1980, 315, referenced to Thorpe (1956).

3 Deleuze pursued the connection between music and politics in his essay *Périclès et Verdi* (1988a) – Pericles the governor of Athens, who rebuilt the Acropolis in its classic age, and Verdi the great composer of operas.

4 Schlegel and Goethe are often credited with the expression, which they certainly repeated (Pascha, 2004).
5 Nietzsche (1888) *The Case of Wagner*. [Deleuze's footnote].
6 See Detienne (1989) pp. 51–2. [Deleuze's footnote].
7 Concerning the eternal return, Zarathustra asks his own animals: 'Have you already made a hurdy-gurdy song of this?' Nietzsche (1883) third part, 'The Convalescent', 330. [Deleuze's footnote].
8 See the different stanzas of 'The Seven Seals' in Nietzsche (1883) 340–3. [Deleuze's footnote]
9 On the question of 'sanctuary', that is, of God's territory, see Jeanmaire (1970) 193. 'One encounters him everywhere, yet he is nowhere at home. . . . He insinuates himself more than he imposes himself.' [Deleuze's footnote].
10 Eugène Dupréel elaborated a set of original notions, 'consistency' (in relation to 'precariousness'), 'consolidation', 'interval', 'intercalation'. See Dupréel (1933, 1939, 1961); Bachelard, in *La dialectique de la durée*, draws on Dupréel. [Deleuze and Guattari's footnote].
11 Woolf (1980) vol. 3, 209. [Brian Massumi's footnote].

4 Façade and Landscape

1 The translation here is made directly from the text quoted by Deleuze and Guattari on p. 212 of the French original of *Mille Plateaux*. It is informed by two published translations: Chrétien, 1181–90a, 432–3 ; and Chrétien, 1181–90b, 88–89. The version in Brian Massumi's translation of Deleuze's and Guattari's text is from the latter source.
2 Jakob von Uexküll's work is cited by Heidegger (1929–30) and Merleau-Ponty (1995) as well as by Deleuze and Guattari.

5 City and Environment

1 From all these standpoints, François Châtelet questions the classical notion of the city-state and doubts that the Athenian city can be equated with any variety of State: 'La Grèce classique, la Raison, l'Etat', (in Rosa,1978). Islam was to confront analogous problems, as would Italy, Germany, and Flanders

beginning in the eleventh century; in these cases, political power does not imply the State-form. An example is the community of Hanseatic towns, which lacked functionaries, an army, and even legal status. The town is always inside a network of towns, but, precisely, 'the network of towns' does not always coincide with 'mosaic of States'. On all of these points, see Fourquet and Murard, 1976, 79–106. [Deleuze and Guattari's note, 1980 565–6].

2 See Gregory Bateson in Ballantyne, 2005, 35–7, 71–2, 109–10.
3 This passage is discussed in Diane Ghirardo, 'The Architecture of Deceit' in Balllantyne, 2002, 63–71; and in Johnson, 2001, 36–8.

Bibliography

Adam, Peter (2000) *Eileen Gray: Architect/Designer*, New York: Abrams 1987, revised 2000.

Arnold, Dana and Ballantyne, Andrew (2004) *Architecture as Experience: Radical Change in Spatial Practice*, London: Routledge.

Artaud, Antonin (1947) *Pour en finir avec le jugement de dieu,* translated by Helen Weaver (1988) 'To Have Done with the Judgement of God', in Antonin Artaud, *Selected Writings,* edited by Susan Sontag, New York: Farrar, Strauss & Giroux.

Bachelard, Gaston (1950) *La dialectique de la durée,* Paris: Presses Universitaires de France.

Ballantyne, Andrew (1997) *Architecture, Landscape and Liberty: Richard Payne Knight and the Picturesque*, Cambridge: Cambridge University Press.

—— (2002) *What is Architecture?,* London: Routledge.

—— (2005) *Architecture Theory*, London: Continuum.

Bataille, Georges (1931) *L'Anus solaire*, illustrated by André Masson, Paris: André Simon; text in (1970) *Oeuvres complètes*, vol. 1, Paris: Gallimard.

—— (1949) *La Part maudite: essai d'économie générale*, 3 vols, Paris: Editions de Minuit; translated by R. Hurley (1988, 1991) *The Accursed Share*, 2 vols (the title of vol. 2 is *The Accursed Share: Volumes 2 and 3*) New York: Zone Books.

Bell, Vikki (1999) *Performativity and Belonging*, London: Sage.

Berendt, John (1994) *Midnight in the Garden of Good and Evil: a Savannah Story*, New York: Random House.

Buchanan, Ian (2000) *Deleuzism: A Metacommentary*, Edinburgh: Edinburgh University Press.

—— (2005) 'Space in the Age of Non-Place', in Buchanan and Lambert (2005) 16–35.

Buchanan, Ian and Lambert, Gregg (2005) *Deleuze and Space*, Edinburgh: Edinburgh University Press.

Büchner, Georg (1839) *Lenz*, Frankfurt: Deutscher Klassiker Verlag [1999]; translated by Richard Sieburth (2004) *Lenz*, New Yok: Archipelago Books.
—— (1993) translated by John Reddick, *Complete Plays,* Lenz *and Other Writings*, Harmondsworth: Penguin.
Butler, Samuel (1872) *Erewhon*, London.
Byron, [George Gordon] Lord (1812–18) *Childe Harold's Pilgrimage*, London: John Murray.
Cache, Bernard (1995) translated by Anne Boyman, *Earth Moves: The Furnishing of Territories*, Cambridge, MA: MIT Press; French edition (1997) *Terre meuble,* Orleans: Editions HYX.
Camazine, Scott, *et al.* (2001) *Self-Organization in Biological Systems*, edited by Scott Camazine, Jean-Louis Deneubourg, Nigel R. Franks, James Sneyd, Guy Theraulaz, Eric Bonabeau, Princeton, NJ: Princeton University Press.
Canetti, Elias (1973) translated by Carol Stewart, *Crowds and Power*, Harmondsworth: Penguin.
Chipperfield, David (1994) *Theoretical Practice*, London: Artemis.
Chrétien de Troyes (1181–90a) *Perceval (Le Conte du Graal)* translated by William W. Kibler (2004) 'The Story of the Grail (Perceval)' in *Arthurian Romances,* Harmondsworth: Penguin.
—— (1181–90b) *Perceval (Le Conte du Graal)* translated by Robert White Linker (1952) *The Story of the Grail*, Chapel Hill, NC: University of North Carolina Press.
Clare, John (1819) 'The Woodman', in (1984) *John Clare: A Critical Edition of the Major Works*, edited by Eric Robinson and David Powell, Oxford: Oxford University Press.
Colebrook, Claire (2002) *Gilles Deleuze*, London: Routledge
—— (2006) *Deleuze: A Guide for the Perplexed,* London: Continuum.
Cressole, Michel (1973) *Deleuze*, Paris: Editions Universitaires.
Dawkins, Richard (1976) *The Selfish Gene*, Oxford: Oxford University Press.
Delanda [or De Landa], Manuel (1991) *War in the Age of Intellligent Machines*, New York: Zone Books.
—— (1997) *A Thousand Years of Nonlinear History*, New York: Zone Books.
—— (2002) *Intensive Science and Virtual Philosophy,* London: Continuum.
—— (2005) 'Space: Extensive and Intensive, Actual and Virtual', in Buchanan and Lambert (2005) 80–8.

—— (2006) *A New Philosophy of Society: Assemblage Theory and sociela complexity*, London: Continuum.

Delbos, Victor (1893) *Le Problème moral dans la philosophie de Spinoza et dans l'histoire du spinozisme*, Paris: Alcan.

—— (1950) *Le Spinozisme*, Paris: Vrin.

Deleuze, Gilles (1953) *Empirisme et subjectivité: essai sur la nature humaine selon Hume*, Paris: Presses Universitaires de France; translated by Constantin V. Boundas (1991) *Empiricism and Subjectivity: An Essay on Hume's Theory of Human Nature*, New York: Columbia University Press.

—— (1962) *Nietzsche et philosophie*, Paris: Presses Universitaires de France; translated by Hugh Tomlinson (1983) *Nietzsche and Philosophy*, London: Athlone.

—— (1963) *La Philosophie critique de Kant: doctrine des facultés*, Paris: Presses Universitaires de France; translated by Hugh Tomlinson and Barbara Habberjam (1984) *Kant's Critical Philosophy: The Doctrine of the Faculties*, Minneapolis, MN: Minnesota University Press.

—— (1965) *Nietzsche*, Paris: Presses Universitaires de France.

—— (1966) *Bergsonisme*, Paris: Presses Universitaires de France; translated by Hugh Tomlinson and Barbara Habberjam (1988) *Bergsonism*, New York: Zone Books.

—— (1968) *Différence et répétition*, Paris: Presses Universitaires de France; translated by Paul Patton (1994) *Difference and Repetition*, London: Athlone.

—— (1968) *Spinoza et le problème d'expression*, Paris: Les Editions de Minuit; translated by Martin Jouchin (1990) *Expressionism in Philosophy: Spinoza*, New York: Zone Books.

—— (1969) *Logique du sens*, Paris: Editions du Minuit; translated by Mark Lester and Charles Stivale (1990) *Logic of Sense*, edited by Constantin V. Boundas, New York: Columbia University Press.

—— (1970) *Spinoza: Philosophie pratique*, Paris: Presses Universitaires de France; revised and expanded (1981) Paris: Les Edition de Minuit; translated by Robert Hurley (1988) *Spinoza: Practical Philosophy*, San Francisco, CA: City Lights.

—— (1983) *Cinéma 1: L'Image-mouvement,* Paris: Les Editions de Minuit; translated by Hugh Tomlinson and Barbara Habberjam (1986) *Cinema 1: The Movement-Image*, London: Athlone.

—— (1985) *Cinéma 2: L'Image-temps,* Paris: Les Editions de Minuit; translated by Hugh Tomlinson and Robert Galeta (1989) *Cinema 2: The Time-Image*, London: Athlone.

—— (1988a) *Périclès and Verdi: la philosophie de François-Châteler*, Paris, Minuit.

—— (1988b) *Le Pli: Leibniz et le baroque,* Paris: Les Editions de Minuit; translated by Tom Conley (1993) *The Fold: Leibniz and the Baroque*, London: Athlone.

—— (1990) *Pourparlers*, Paris: Les Editions de Minuit; translated by Martin Jouchin (1995) *Negotiations*, New York: Columbia University Press.

—— (1993) *Critique et clinique*, Paris: Les Editions de Minuit; translated by Martin Jouchin (1997) *Essays Critical and Clinical*, Minneapolis, MN: Minnesota University Press.

—— (2002) *Iles désertes*, edited by David Lapoujade, Paris: Editions de Minuit; translated by Michael Taormina (2004) *Desert Islands*, New York: Semiotext(e).

—— (2003) *Deux régimes de fous*, edited by David Lapoujade, Paris: Editions de Minuit; translated by Ames Hodges and Michael Taormina (2006) *Two Regimes of Madness*, New York: Semiotext(e).

Deleuze, Gilles and Guattari, Félix (1972) *Capitalisme et schizophrénie 1: L'Anti-Oedipe*, Paris: Editions du Minuit; translated by Robert Hurley, Mark Seem and Helen R. Lane (1977) *Capitalism and Schizophrenia 1: Anti-Oedipus*, New York: Viking.

—— (1975) *Kafka: Pour une litérature mineure*, Paris: Editions du Minuit; translated by Dana Polan (1986) *Kafka: Toward a Minor Literature*, Minneapolis, MN: Minnesota University Press.

—— (1976) *Rhizome*, Paris: Editions du Minuit.

—— (1980) *Capitalisme et schizophrénie 2: Mille plateaux;* translated by Brian Massumi (1987) *Capitalism and Schizophrenia 2: A Thousand Plateaus*, Minneapolis, MN: Minnesota University Press.

—— (1991) *Qu'est-ce que la philosophie?*, Paris: Editions du Minuit; translated by Graham Burchell and Hugh Tomlinson (1994) *What is Philosophy?*, New York: Columbia University Press.

Deleuze, Gilles and Parnet, Claire (1977) *Dialogues,* Paris, Flammarion; translated by Hugh Tomlinson and Barbara Habberjam (1987) *Dialogues,*

London: Athlone; reissued with supplementary material (2002) *Dialogues II*, London: Continuum.
Detienne, Marcel (1989) translated by Arthur Goldhammer (1989) *Dionysus at Large*, Cambridge, MA: Harvard University Press.
Diogenes of Sinope (*c.* 340 BC) translated by Guy Davenport (1979) *Herakeitos and Diogenes*, San Francisco, CA: Grey Fox.
Drexler, K. Eric (1986) *Engines of Creation: The Coming Era of Nanotechnology*, New York: Randon House.
Due, Reidar (2007) *Deleuze*, London: Polity.
Dupréel, Eugène (1933) *Théorie de la consolidation: La cause et l'intervalle*, Brussels: M. Lamertin.
—— (1939) *Esquisse d'une philosophie des valeurs*, Paris: Alcan.
—— (1961) *La consistence et la probabilité objective*, Brussels: Académie Royale de Belgique.
Engels, Friedrich (1845) *Die Lagen der arbeitenden Klasse in England*, Leipzig, translated by Florence Wischnewetsky, *The Condition of the Working Classes in England in 1844*, reprint 1973, Moscow: Progress.
Feher, Michel (1989) *Zone: Fragments for a History of the Human Body*, edited by Michel Feher, 3 vols (numbered 3, 4 and 5), Cambridge, MA: MIT Press.
Fitton, R.S. (1989) *The Arkwrights: Spinners of Fortune*, Manchester: Manchester University Press.
Foucault, Michel (1970) '*Theatrum Philosophicum*', in *Critique* 282, 885–908, Paris; translated by Sherry Simon (1977) '*Theatrum Philosophicum*', in *Language, Counter-Memory, Practice*, edited by Donald F. Bouchard, Ithaca: Cornell University Press.
Fourquet, François and Murard, Lion (1976) *Les équipements de pouvoir: ville, territories et équipements collectifs*, Paris: 10/18.
Freud, Sigmund (1911) 'Psychoanalytische Bemerkungen Über Einen Autobiographisch Beschiebenen Fall Von Paranoia (*Dementia Paranoides*)' Jb. psychoanalyt. Psychopath. Forsch., 3 (1) 9–68; translated by James Strachey and Angela Richards (1955) 'Psycho-Analytic Notes upon an Autobiographical Account of a Case of Paranoia (*Dementia Paranoides*) (Schreber)', in Sigmund Freud (1991), *Case Studies 2* (vol. 9 of *The Penguin Freud Library*) Harmondsworth: Penguin.
Fuller, Buckminster (1963) *Operating Manual for Spaceship Earth*, New York: E.P. Dutton.

Genosko, Gary (2002) *Félix Guattari: An Aberrant Introduction*, London: Continuum.

—— (2006) 'Busted: Félix Guattari and the *Grande Encyclopédie des Homosexualités*' in *Rhizomes*, 11/12, Fall 2005/Spring 2006.

Gilliard, E.T. (1969) *Birds of Paradise and Bower Birds*, London: Weidenfeld.

Goodman, Nelson (1978) *Ways of Worldmaking*, Hassocks: Harvester Press.

Guattari, Félix (1979) *L'Inconscient machnique: essays de schizo-analyse*, Clamecy: Editions Recherches.

—— (1984) translated by Rosemary Sheed, *Molecular Revolution: Psychiatry and Politics*, edited by Ann Scott, Harmondsworth: Penguin.

—— (1989) *Les troìs écologies*, Paris: Galilée, translated by Ian Pindar and Paul Sutton (2000) *The Three Ecologies*, London: Athlone.

—— (1992) *Chaosmose*, Paris: Galilée, translated by Paul Bains and Julian Pefamis (1995) *Chaosmosis: An Eco-Aesthetic Paradigm*, Sydney: Power Publications.

—— (1996a) translated by David L. Sweet and Chet Wiener, *Soft Subversions*, edited by Sylvère Lotringer, New York: Semiotext(e).

—— (1996b) *The Guattari Reader*, edited by Gary Genosko, Oxford: Blackwell.

—— (2002) *'La Philosophie est essentielle à l'existence humaine': entretien avec Antoine Spire*, Paris: L'Aube.

—— (2005) *Ecrits pour l'Anti-Oedipe*, edited by Stéphane Nadaud, Paris: Léo Scheer; translated by Kélina Gotman (2006) *The Anti-Oedipus Papers*, New York: Semiotext(e).

Harris, Paul André (2005) 'To See with the Mind and Think Through the Eye: Deleuze, Folding Architecture, and Simon Rodia's Watts Towers', in Buchanan and Lambert (2005) 36–60.

Haudricourt, André (1962) 'Domestication des animaux, culture des plantes et traitement d'aurtui', in *L'Homme*, vol. 2, no. 1 (January–April 1964) 40–50.

—— (1964) 'Nature et culture dans la civilisation de l'igname: l'origine des cloues et des dans, *L'Homme*, vol. 4, no. 2 (January–April 1964) 93–104.

Heidegger, Martin (1929–30) [1983] *Die Grundbegriffe der Metaphysik. Welt – Endlichkeit – Einsamkeit*, Frankfurt: Vittorio Klostermann; translated by William McNeill and Nicholas Walker (1995) *The Fundamental Concepts of Metaphysics: World, Finitude, Solitude*, Bloomington, IN: Indiana University Press.

Hensel, Michael (2004) *Emergence: Morphogenetic Design Strategies*, edited by Michael Hensel, Achim Menges and Michael Weinstock, London: Wiley-Academy.

—— (2006) *Techniques and Technologies in Morphogenetic Design*, edited by Michael Hensel, Achim Menges and Michael Weinstock, London: Wiley-Academy.

Hofstadter, Douglas R. (1979) *Gödel, Escher, Bach: An Eternal Golden Braid,* New York: Basic Books.

Holland, Eugene W. (1999) *Deleuze and Guattari's* Anti-Oedipus*: Introduction to Schizoanalysis*, London: Routledge.

Horrobin, David (2001) *The Madness of Adam and Eve: How Schizophrenia Shaped Humanity*, London: Transworld.

Hume, David (1739) *A Treatise of Human Nature*, Edinburgh; edited by L.A. Selby-Bigge and P.H. Nidditch (1978) Oxford: Clarendon Press.

—— (1751) *An Enquiry Concerning Human Understanding*, Edinburgh; edited by L.A. Selby-Bigge and P.H. Nidditch along with *An Enquiry Concerning the Principles of Morals* and *A Dialogue* (1975) Oxford: Clarendon Press.

—— (1779) *Dialogues Concerning Natural Religion*, London.

—— (1777) *Essays and Treatises on Several Subjects*, 2 vol., London: T. Cadell

Jaeglé, Claude (2005) *Portrait oratoire de Gilles Deleuze aux yeux jaunes*, Paris: Presses Universitaires de France.

James, Henry (1909) 'Preface', in *The Wings of the Dove* (first published 1902) New York: Scribner.

Jeanmaire, Henri (1970) *Dionysus, histoire du culte de Bacchus*, Paris: Payot.

Johnson, Steven (2001) *Emergence: The Connected Lives of Ants, Brains, Cities and Software*, New York: Scribner.

Kaufmann, Eleanor (2001) *The Delirium of Praise: Bataille, Blanchot, Deleuze, Foucault, Klossowski*, Baltimore, MD: Johns Hopkins University Press.

Khalfa, Jean (2003) *An Introduction to the Philosophy of Gilles Deleuze*, London: Continuum.

Lavin, Sylvia (1992) *Quatremère de Quincy and the Invention of a Modern Language of Architecture*, Cambridge, MA: MIT.

Lecercle, Jean-Jacques (1985) *Philosophy Through the Looking Glass: Language, Nonsense, Desire*, La Salle, IL: Open Court.

Le Corbusier (1923) *Vers une architecture*, Paris: Crès; translated by Frederick Etchells (1987) *Towards a New Architecture*, London, Architectural Press.

Leroi-Gourhan, André (1945) *Milieu et techniques*, Paris: Albin Michel.

—— (1964) *Le Geste et la parole*, Paris: Albin Michel; translated by Anna Bostock Berger (1993) *Gesture and Speech*, Cambridge, MA: MIT Press.

Lestel, Dominique (2001) *Les Origines animales de la culture*, Paris: Flammarion

Loughlin, Gerard (2003) *Alien Sex: The Body and Desire in Cinema and Theology*, Oxford: Blackwell.

Lowry, Malcolm (1933) *Ultramarine*, Philadelphia, PA: Lippincott [1962].

Lynn, Greg (1993) *Folding in Architecture*, London: Academy. Revised edition 2004.

—— (1998a) *Folds, Bodies and Blobs*, Brussels: La Lettre Volée.

—— (1998b) *Animate Form*, Princeton, NJ: Princeton Architectural Press.

—— (2006) *Predator,* Seoul: DAMDI Publishing.

Malamud, Bernard (1966) *The Fixer,* New York: Farrar, Strauss & Giroux.

Marks, John (1998) *Deleuze: Vitalism and Multiplicity*, London: Pluto.

Marshall, Alan John (1954) *Bower Birds*, Oxford: Clarendon Press.

Marx, Karl and Engels, Friedrich (1848) *Manifest der Kommunistischen Partei*, London: Bildungsgesellschaft für Arbeiter; translated by Samuel Moore (1888) *The Communist Manifesto*, reprinted Harmondsworth: Penguin, 1967.

Marx, Karl (1867–94) *Das Kapital: Kritik der politischen Ökonomie*, Hamburg: Meissner, translated by Ben Fowkes (1976) *Capital: A Critique of Political Economy*, Harmondsworth: Penguin, 3 vols.

Massumi, Brian (1992) *A User's Guide to Capitalism and Schizophrenia*, Cambridge, MA: MIT Press.

—— (2002) *Parables for the Virtual: Movement, Affect, Sensation*, Durham, NC: Duke University Press.

May, Todd (2001) *Our Practices, Our Selves, or: What it Means to be Human*, University Park, PA: Penn State Press.

—— (2005) *Gilles Deleuze: An Introduction*, Cambridge: Cambridge University Press.

Melville, Herman (1851) *Moby-Dick, or: The Whale*, New York: Harper & Brothers.

Merleau-Ponty, Maurice (1995) *La Nature: Notes cours du Collège de France,* edited by D. Seglard, Paris: Seuil; translated by R. Vallier (2003) *Nature:*

Course Notes from the Collège de France, Evanston, IL: Northwest University Press.
Minsky, Marvin (1985) *The Society of Mind*, New York: Simon & Schuster.
—— (2006) *The Emotion Machine: Commonsense Thinking, Artificial Intelligence, and the Future of the Human Mind*, New York: Simon & Schuster.
Nietzsche, Friedrich (1878) *Menschliches Allzumenschliches*; translated by R.J. Hollingdale (1986) *Human, All Too Human,* Cambridge: Cambridge University Press.
—— (1883) *Also sprach Zarathustra:ein Buch für Alle und Keinen*, translated by Walter Kaufmann (1954) *Thus Spoke Zarathustra: a Book for All and None,* in *The Portable Nietzsche*, New York: Viking.
—— (1886) *Jenseits von Gut und Böse – Vorspiel einer Philosophie der Zukunft*; translated by R.J. Hollingdale (1973) *Beyond Good and Evil: Prelude to a Philosophy of the Future,* Harmondsworth: Penguin.
—— (1887) *Zur Genealogie der Moral – Eine Streitschrift*; translated by Walter Kaufmann and R.J. Hollingdale (1967) *The Genealogy of Morals: A Polemic*, New York: Random House.
—— (1888) translated by Walter Kaufmann (1967) 'The Case of Wagner', in *The Birth of Tragedy and The Case of Wagner*, New York: Vintage.
Oberlin, Johann Friedrich (1778) translated by Richard Sieburth (2004) 'Mr. L . . .', in Büchner (1839) 81–127.
Orlean, Susan (1998) *The Orchid Thief*, New York: Random House.
Pascha, Khaled Saleh (2004) *'Gefrorene Musik': Das Verhältnis von Architektur und Musik in der ästhetischen Theorie*, Berlin: unpublished PhD thesis.
Parr, Adrian (2005) *The Deleuze Dictionary*, edited by Adrian Parr, Edinburgh: Edinburgh University Press.
Protevi, John (2001) *Polticial Physics: Deleuze, Derrida and the Body Politic*, London: Athlone.
Proust, Marcel (1913–27) *A la recherche de temps perdu*, Paris, Grasset; translated by S. Moncrief, A. Mayor and T. Kilmartin, revised by D.J. Enright (1992) *In Search of Lost Time*, 6 vols, London: Chatto & Windus.
Rajchman, John (1998) *Constructions*, Cambridge, MA: MIT Press.
—— (2000) *The Deleuze Connections*, Cambridge, MA: MIT Press.

Reiser, Jesse and Umemoto, Nanako (2006) *Atlas of Novel Tectonics*, Princeton, NJ: Princeton University Press.

Riesman, David (1950) *The Lonely Crowd* (revised edition 1961) New Haven, CT: Yale University Press.

Rosa, Alberto Asor, Chatelet, François, Dadoun, Roger, Delacampagne, Christian, *et al.* (1978) *En marge. L'Occident et ses 'autres'*, Paris: Aubier Montaigne.

Ruskin, John (1862) 'Ad Valorem', in *Unto This Last: Four Essays on the First Principles of Political Economy*, London; collected in (1985) *Unto This Last and Other Writings*, edited by Clive Wilmer, Harmondsworth: Penguin.

Rykwert, Joseph (1996) *The Dancing Column*, Cambridge, MA: MIT Press.

Sasso, Robert and Villani, Arnaud (2003) *Le vocabulaire de Gilles Deleuze*, Paris: Centre de Recherches d'Histoire des Idées.

Schrader, Paul (1972) *Transcendental Style in Film: Ozu, Bresson, Dreyer*, Berkeley, CA: University of California Press.

Schreber, Daniel Paul (1903) *Denkwürdigkeiten eines Nervenkranken*, Leipzig; translated by I. Macalpine and T.A. Hunter (1955) *Memoirs of my Nervous Illness*, London.

Richard Sieburth (2004) 'Translator's Afterword', in *Büchner* (1839) 165–197.

Simondon, Gilbert (1958) *Du mode d'existence des objets techniques*, Paris: Aubier.

—— (1964) *L'Individu et sa genèse physico-biologique*, Paris: Presses Universitaires de France.

Smith, Adam (1776) *An Inquiry into the Nature and Causes of the Wealth of Nations*, Edinburgh.

Spinoza, Baruch (1677a) *Tractatus Theologico-Politicus*, Amsterdam; translated by Edwin Curley (1985) 'Theological-Political Treatise', in *The Collected Works of Spinoza*, vol. 1, Princeton, NJ: Princeton University Press.

—— (1677b) *Ethica ordine geometrico demonstrata*, Amsterdam; translated by Samuel Shirley (1992) *Ethics; Treatise on the Emendation of the Intellect; Selected Letters*, Indianapolis, IN: Hackett.

Thorpe, W.H. (1956) *Learning and Instinct in Animals*, London: Methuen.

Uexküll, Jakob von (1934) *Streifzüge durch die Umwellen von Tieren und Menschen*, Hamburg: Rowohlt; translated by Philippe Muller (1965) *Mondes animaux et monde humain,* and *Théorie de la signification*, Paris: Gonthier.

Winckelmann, Johann Joachim (1755) *Gedanken über die Nachahmung der greichischen Werke in der Mahlerey und Bildauer-Kunst*, Dresden; translated by Henry Fuseli (1765) *Reflections on the Painting and Sculpture of the Greeks with Instructions for the Connoisseur, and an Essay on Grace in Works of Art*, London.

Wood, David (2004) 'Territoriality and Identity at RAF Menwith Hill' in *Architectures: Modernism and After*, edited by Andrew Balllantyne, 142–62, Oxford: Blackwell.

Virginia Woolf, *The Diary of Virginia Woolf*, edited by Anne Olivier Bell assisted by Andrew McNeillie, 6 vols (London: The Hogarth Press, 1980) vol. 3: 1925–1930.

Zourabichvili, François (2004) *Le vocabulaire de Deleuze*, Paris: Ellipses.

Index

Numbers in **bold** type indicate illustrations